Electromagnetic Radiations

The book delivers an understanding of emission theory and its effects on different strata of life. It contains seven chapters including probable remedial measures and solutions to increase reduced radiation life expectancy. The text explains important topics such as the compatibility of the human body and wireless communication, applications and effectiveness of radiating power, energy harvesting, green energy solutions, and the human nervous system.

This book:

- Discusses topics related to radiation and electromagnetic emissions, including their sources, effects, and ways to reduce exposure.
- Covers various aspects of the impact of electromagnetic fields on health and the environment, including measurement and modeling techniques, exposure assessment, and health effects.
- Explains electromagnetic emissions and their applications, as well as the impact of radiation on living organisms, including flora, fauna, and human beings.
- Provides a detailed analysis of the effects of radiation on animal and plant life.
- Highlights the potential benefits of electromagnetic emissions and provides information on how to mitigate the negative effects of radiation.

It is primarily written for senior undergraduates, graduate students, and academic researchers in the fields including electrical engineering, electronics, communications engineering, and physics.

Electromagnetic Radiations

Exposure and Impact

Prutha Prashant Kulkarni
and Parikshit N. Mahalle

CRC Press
Taylor & Francis Group
Boca Raton London New York

CRC Press is an imprint of the
Taylor & Francis Group, an **informa** business

First edition published 2025
by CRC Press
2385 NW Executive Center Drive, Suite 320, Boca Raton FL 33431

and by CRC Press
4 Park Square, Milton Park, Abingdon, Oxon, OX14 4RN

CRC Press is an imprint of Taylor & Francis Group, LLC

ISBN: 978-1-032-71566-7 (hbk)
ISBN: 978-1-032-96799-8 (pbk)
ISBN: 978-1-003-59071-2 (ebk)

DOI: 10.1201/9781003590712

Typeset in Sabon
by Deanta Global Publishing Services, Chennai, India

Contents

About the authors

Prutha Prashant Kulkarni, with over a decade of research experience, specializes in antenna miniaturization and metamaterials for wireless applications. Her research primarily focuses on metamaterial-inspired antenna designs and miniaturized antennas for GPS, CubeSat, IoT, and SATCOM applications. Dr Kulkarni holds four patents and has authored over ten papers in prestigious journals and conferences. She is an active member of IEEE AP-S and MTT-S societies and serves as Vice-Chair of the IEEE APS/MTT/EMC Joint Chapter, Pune Section. Additionally, she has delivered several invited talks, including "Navigating the Electromagnetic Spectrum: Understanding Hazards, Detoxing Digitally, and Balancing Connectivity," "Radiation Exposures: A New Normal," "Antennas and Metamaterials," and "Antennas for IoT." Dr Kulkarni holds a Ph.D. in Information and Communication Engineering, specializing in the miniaturization of LP and CP antennas for GNSS applications. She has authored several books, including *Antennas for IoT* (Artech House, 2023), and received the IEEE Pune Section "Women in Engineering of the Year" award in 2024.

Parikshit N. Mahalle is a senior member of IEEE and serves as the Professor and Dean of Research and Development at Vishwakarma Institute of Technology, Pune, India. Prior to this, he worked as the Dean of Research and Development and Head of the Department of Artificial Intelligence and Data Science at Vishwakarma Institute of Information Technology, Pune, India. He also served as Professor and Head of the Department of Computer Engineering at Sinhgad Institutes. He completed his PhD from Aalborg University, Denmark, and completed his Postdoctoral Research at CMI, Copenhagen, Denmark. He has 24 years of teaching and research experience. He is an ex-member of the Board of Studies in Computer Engineering, an ex-chairman of Information Technology at Savitribai Phule Pune University, an ex-chairman of the Board of Studies – AI&Ds at VIIT, and a member of the BoS and Academic Council at more than 30 Universities and autonomous colleges across India. He holds 52 patents and has authored 400+ research publications (Google Scholar citations: 4000+, H index: 28; Scopus citations: 2000+, H index: 21; and Web of Science citations: 545,

H index: 11). He has also authored/edited 67 books with publishers such as Springer, CRC Press, Cambridge University Press, etc. He is the Editor-in-Chief of the *International Journal of Rough Sets and Data Analysis* (IGI Global) and *Research Journal of Computer Systems and Engineering (RJCSE)* and Associate Editor of the *Journal of Affective Computing and Human Interfaces (JACHI)* (IGI Global). Additionally, he serves as a member of the Editorial Review Board for the *International Journal of Ambient Computing and Intelligence* (IGI Global) and as a reviewer for various prestigious transactions, journals, and conferences. His research interests include machine learning, data science, algorithms, the Internet of Things, identity management, and security. He is currently guiding eight PhD students in the fields of IoT and machine learning; seven students have successfully defended their PhDs under his supervision from SPPU, and two students have completed postdoctoral research under his mentorship from NTU, Taiwan. He is also the recipient of the "Best Faculty Award" by Sinhgad Institutes and Cognizant Technology Solutions, the International Level S4DS Distinguished Researcher of the Year 2023, the State Level Meritorious Teacher Award, and the Distinguished Research Guide Award at IEEE ICTBIG 2024, organized by the Symbiosis University of Applied Sciences (SUAS), Indore. He has delivered 400-plus lectures at national and international levels. His book *Design and Analysis of Algorithms* is referred as a textbook in IITs and NITs, and his book *Data Analysis on Pandemic* (CRC Press) received two international awards in 2020. His edited book, *Data Science: Techniques and Intelligent Applications*, has been awarded the prestigious Choice Outstanding Academic Titles Award for 2024. He is also a certified ISO 27001:2022 Lead Auditor. He has been an invited guest faculty member at several international universities, including UMA in Lima, Peru, South America, and National Taipei University in Taiwan.

Chapter 1

Introduction

EVOLUTION OF LIFE FULL OF GADGETS

In a modern world that is always changing, electronic devices and appliances are absolutely necessary; with the support of science and technology, anything is possible. Because of this, we come into contact with a wide array of electronic devices in our day-to-day lives without even being aware of them. The consequence of this is that we are nearly wholly reliant on them.

There are many different kinds of electronic gadgets that fall under the category of devices. The human experience is made more accessible and innovative as a result of their actions in this manner. The market is rich with a large assortment of smart equipment, which is true regardless of our location. There is a significant majority of those that are quite useful and productive. The development of more advanced technologies has made it easier to make use of a wide range of portable electronic gadgets. Everything from the alarm clock in the morning to the lights at night is something that we all rely on. In this society, machines are the most powerful force in comparison to both human activity and the activity brought about by machines. Because humans are intrinsically dependent on them and because they are the key to all human pursuits, there is no discipline that operates entirely with humans. A number of electronic items, including laptops, smartphones, home appliances, and smartwatches, are examples of things that we simply cannot imagine living without. The employment of various gadgets has resulted in the resolution of a great number of problems that are related to people who have physical disabilities. We are unable to survive without the utilization of technology in our day-to-day lives since it has had a tremendous impact on us in a variety of different ways.

The advent of various technological devices has enabled us to live lives that are both more joyful and more comfortable. During this transition from summer to winter, we are making use of both air conditioning and room heaters. Notable among the extra equipment are a blow dryer and a straightener for the hair. The absence of these gadgets makes it hard to identify any house that may be present. Various devices have the ability

DOI: 10.1201/9781003590712-1

to save a considerable quantity of space. The use of telephones in the past required users to remain still when chatting; however, the advent of smartphones has made it possible for anyone to engage in conversation regardless of their location. From the beginning, our preferred devices have been the iPod, the MP3, the PlayStation, and any other electronic gadgets that are designed to provide enjoyment. Technology in its modern form is the only factor that can account for the occurrence of these events. The employment of these devices helps to cultivate a sense of joy and camaraderie among members of the family. Because of their capacity to alleviate monotony and loneliness in our day-to-day lives, they have acquired the status of a vital resource. When used as an addiction, technology may yield positive results, but they also have the potential to have negative consequences. As a result, it is feasible to set a time limit in order to prevent addiction from occurring.

According to the findings of online polls on social media platforms, 29% of youngsters are fluent in the use of modern technology, and by the time they reach primary school, 70% of them have mastered it. Children who use electronic gadgets on a regular basis are more likely to have a range of negative outcomes, such as anxiety, trouble learning, attention deficiencies, and loss of focus. As a result, it is of the utmost importance that we realize the reality that technology and smartphones make our day-to-day life easier. Since this is the case, it is of the utmost importance that we do not allow ourselves to get dependent on them. There is a tendency among people in today's society to possess a multitude of electronic devices and to become fixated on them, which is not beneficial to their health. Because the brightness of the screen leads us to feel uncomfortable, we are unable to exercise any control over the situation and are therefore compelled to stare at the screen. This is not useful. In light of this, the realization that if we make constructive use of technology, we will all be in a favorable position is helpful. On the other hand, if we utilize it in a manner that is not positive, it should not come as a surprise that we are in difficulty.

The development of technological devices has profoundly impacted every aspect of our daily lives, seamlessly integrating into our routines and enhancing both productivity and communication. Whether it's smartwatches monitoring our health or smartphones facilitating communication, these gadgets have become indispensable. It's difficult to imagine navigating the complexities of modern life without them. Today, businesses rely heavily on specialized software, gadgets, and computers to make informed decisions, manage data effectively, and streamline operations. The healthcare industry, educational systems, and transportation sectors have all witnessed significant advancements due to these technologies. The widespread availability of electronic devices has democratized access to information and entertainment, transforming how we interact with content through e-books, streaming services, and social media platforms.

In educational institutions, students attest to the crucial role of technological devices in their academic success. These tools provide access to vast amounts of knowledge, enhance communication and collaboration, and make learning more engaging and effective. Whether using tablets, e-books, or educational software, students benefit from a richer and more interactive educational experience. Our work environments and lifestyles have been transformed by this technological revolution. As we transition into an era of increased connectivity and opportunities, these devices have become virtually indispensable. Without them, many individuals find it challenging to leave their homes and remain productive. Looking ahead, technological advancements will continue to solidify the importance of these devices in our lives, further enhancing their roles and capabilities. The future promises a plethora of developments that will continue to shape and improve our daily existence, underscoring the irreplaceable value of technology in our world.

Gadgets: from a merely convenient object to an essential instrument of contemporary life

As a result of the incorporation of various electronic devices into our everyday lives, the ways in which we work, communicate, and entertain ourselves have been completely transformed. Gadgets [1] have evolved from being merely convenient to becoming vital tools that enhance our productivity, connectivity, and overall experience. This includes smartphones and tablets, which now enable users to access Bizzo Casino, as well as smartwatches and virtual assistants.

The use of various electronic devices has brought about a change in the way we communicate with one another. Regardless of where we are physically located, the process of keeping communication with friends, family, and associates has become much simpler as a result of the arrival of smartphones. With the help of these portable gadgets, we have the ability to make phone calls, send text messages, and access social media platforms at the tip of our fingers. Because of the proliferation of messaging programs and video conferencing services, it is now feasible to have real-time conversations with people located in different parts of the world. Gadgets have made communication more accessible, convenient, and efficient in ways that we never imagined were conceivable. This has enabled us to interact with other people in ways that we never thought were possible.

Increasing productivity and flexibility in the workplace through transformation process

Gadgets [2] have not only made communication easier, but they have also completely altered the way in which we conduct our daily activities. Because of mobile devices such as smartphones, tablets, and laptops, we are now able to carry our work with us wherever we go. The ability to remotely access communications, create documents, and collaborate with coworkers is something that we can facilitate. The proliferation of gadgets has led to an increase in our productivity and adaptability, making it possible for us to work while we are on the move and break free from the restrictions of the traditional office setting. Work procedures have been significantly simplified as a result of the combination of cloud storage services and productivity tools, which have made it easier to organize and share information. The employment of various devices has made it possible for us to function in a manner that is more efficient, rapid, and intelligent.

There is an infinite variety of personalized experiences available in the realm of entertainment in the digital age. In addition, the entertainment business has been completely transformed by gadgets. The manner in which we take part in leisure activities and consume media has been revolutionized with the introduction of gaming consoles, electronic readers, and streaming services. Therefore, there is no longer a requirement for tangible media because we now have access to a massive library of movies, television programs, and music that can be accessed on demand. The game experience has been altered by technologies such as virtual reality and augmented reality, which have made it more participatory and immersive. E-readers have not only made it possible for us to read books and articles from any location, but they have also made it possible for us to carry entire libraries in our pockets. We now have access to an infinite number of possibilities and tailored experiences thanks to the proliferation of gadgets, which have opened up new paths for entertainment.

Innovations in health, home automation, and other subjects that are transforming everyday life

In addition, the use of electronic devices has had an impact on a wide range of aspects of our day-to-day lives, such as the automation of our homes and our health and fitness. Smartwatches and fitness sensors make it possible to keep track of our overall health, as well as our sleeping habits and the amount of physical activity we do. They not only provide us with insightful information but also motivate us to adopt better lifestyles. There are a number of home automation gadgets that have improved the effectiveness and convenience of our living environments. Some examples of these devices are voice-activated assistants, smart thermostats, and lighting systems. We can manage and control a range of components of our homes with just a few touches on our smartphones or with a simple voice command. As a result of the utilization of various technological devices, our homes have been turned into intelligent, networked spaces that are tailored to our own preferences and requirements.

But one of the factors that can either unite or disconnect people is the effect on social interactions. Technological devices have not only revolutionized how we communicate but have also changed how we engage with one another socially. However, on the one hand, technology has made it easier to keep connections alive and to keep in touch with loved ones, even when they are physically separated. We are able to instantly share moments, thoughts, and experiences through the use of messaging programs and social media platforms, which help to create connections and bridge gaps between people. On the other hand, the constant presence of electronic gadgets can also make face-to-face encounters more difficult and give the impression of separation. It is of the utmost importance to keep a balanced approach and make certain that the utilization of technology does not hinder true human connections but rather adds to them.

Increasing the quality of educational experiences: the development of education

There has also been a big impact in the sector of education with is the implementation of interactive learning tools [3] and tablets in schools. Students have access to a wide range of instructional resources, which include interactive learning experiences and platforms for collaborative work, thanks to these gadgets. Devices have brought about a revolution in the manner in which knowledge is learned and transmitted by creating an environment that is more conducive to personalized learning and increasing engagement. In addition to conducting scientific experiments and exploring virtual worlds, students have the option to interact with instructional programs that are customized to their individual learning styles and requirements. Gadgets will continue to have an impact on the future of education as technology continues to advance, which will ultimately result in increased autonomy for both students and teachers.

An examination of environmental concerns: striking a balance between convenience and sustainability

It is absolutely necessary to take into consideration the influence that electronics have on the environment as they continue to advance. In addition to being a significant environmental problem, the production and disposal of electronic gadgets also contribute to the accumulation of electronic waste. There are a number of methods by which customers and producers may emphasize sustainability. Some of these approaches include the promotion of environmentally friendly habits, the responsible recycling of obsolete gadgets, and the choosing of items with prolonged lifespans. Additionally, the decrease in our carbon footprint is facilitated by the energy efficiency of the gadgets that we use. By adopting sustainable behaviors and being cognizant of the environmental implications of our technology consumption, we will be able to attain a future that is more sustainable.

In a nutshell, during the course of modern history, electronic devices have progressed from being routine comforts to becoming vital tools for existence. A wide range of facets of our day-to-day life, such as communication, work, entertainment, and education, have been revolutionized as a result of their impact. The advent of various technological devices has resulted in an increase in the level of productivity, pleasure, and connectivity in our lives. In light of the fact that technological progress is proceeding at a breakneck pace, we can anticipate the introduction of new breakthroughs and advancements in the sphere of technology. Wearable technology, the Internet of Things, artificial intelligence, and virtual reality are just a few of the emerging technologies that will continue to have a significant impact on our day-to-day lives and the way we interact with the world. Currently, we are living in a thrilling era in which the potential for growth appears to be unlimited, and technological breakthroughs continue to better our lives in ways that were previously unimagined.

The advantages and disadvantages of gadgets

In the modern period, there is no limit to the technological improvements that have been made. One definition of a gadget is a small mechanism or device that is designed to fulfill a certain function. It might be a separate item, like an espresso machine, or it could be a component of bigger equipment, like a solar battery charger for an automobile. Both of these possibilities are possible scenarios. There is a technological equivalent for almost every work that can be performed in the present day, which is something that is at once impressive and frightening.

Individuals are becoming more slow, despite the fact that technology and gadgets are growing more advanced. There are a range of chores that may be accomplished with the help of these devices without requiring a considerable amount of time or effort on your part. There is a broad variety of sizes, shapes, and purposes that may be found in gadgets. A significant number of them, including games, are completely devoid of any electrical function. The use of calculators and remote controls are two examples of common types of electronic devices.

There has been an increase in our capacity to make consumer gadgets that are both beneficial (like cell phones) and destructive (like electric toys) as a result of technological breakthroughs. In spite of the fact that the vast majority of technological breakthroughs were introduced with the greatest of intentions, not all of them have improved the way people live in the modern era. In this regard, smartphones are among the most prominent instances. Through the promotion of interpersonal ties and the simplification of tasks to the maximum extent feasible, they work toward the goal of simplifying our lives.

However, it also hinders the formation of human relationships, leads to addiction to the internet, and wastes time with its presence. The way in which you use the devices will determine the extent of the benefits and drawbacks associated with technology; nonetheless, your experience will be determined by how you use the gadgets.

For the purpose of boosting students' capabilities and knowledge, gadgets are an essential component of the educational sector. The fact that they need to improve their knowledge and abilities in order to be successful in the future makes it absolutely necessary for them to learn this. The teaching strategies and procedures that teachers use have the potential to be improved. In conclusion, although modern technology is a vital tool for teachers and students, it is not without its fair share of drawbacks. There are billions of people whose lives are made safer, more pleasant, and more wholesome by the availability of gadgets. It is essential for individuals to have access to modern technologies in order to preserve their health and resilience. Our ways of working, playing, shopping, and communicating have all been fundamentally altered by the proliferation of electronic devices.

There are some drawbacks associated with them, despite the fact that they are so pervasive in today's culture. More and more emphasis is being placed on the negative aspects of electronic gadgets, such as the potential for addiction. It is of the utmost importance that every single person who makes use of electronic devices is aware of the impact that these technologies have on both society and our own lives.

Pros and cons of digital era

Facilitates learning that is self-directed

For pupils to have the ability to gain knowledge, it is no longer necessary for parents and teachers to provide aid. It has been proved that the internet is a treasure trove of informative content. Students are able to identify the most amazing assignment writing service in the United Kingdom by utilizing these resources, and they are also able to complete their projects in a rapid fashion. They make use of the internet for their classes, in addition to making use of online resources for research purposes. By utilizing this cutting-edge technology, they have the potential to significantly improve both their expertise and their capabilities.

It is a pleasurable experience to get knowledge

Technology has the potential to make education more accessible to children. It is important to educate your child about the most useful educational

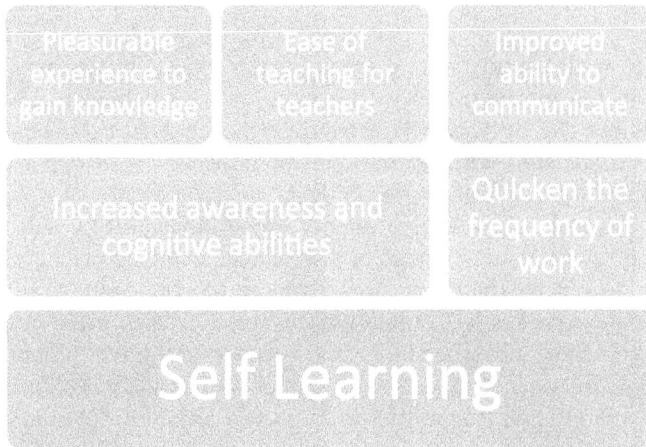

Figure 1.1 The advantages of this digital era.

websites and programs that are relevant to the topics they are interested in. There is a possibility that it will be beneficial for academic and scientific endeavors. To a great extent, youngsters enjoy playing instructional games. Some examples of these games are online tests, tutorials for learning other languages, and tough brainstorming riddles.

Improved ability to communicate

Historically, people communicated with one another through the writing of letters, which could take many days or even months to arrive. Nevertheless, the process has been substantially simplified by the use of email.

Quicken the frequency of the work

A plethora of technological advancements have been developed with the purpose of making day-to-day life easier. Student productivity is increased by the use of technology. Any academic writing task, including a large dissertation, can be completed by a student in a considerably shorter amount of time.

Teachers are able to communicate knowledge to students in a way that is both pleasant and educational

By making use of these resources, educators are now able to create an engaging learning environment for their pupils. Through the utilization of photos, movies, and various other graphic elements, these most recent technologies can be utilized to train students. The creation of an atmosphere that is conducive to learning and the development of an interest in education are both effects of this.

Increased awareness and cognitive abilities

At the end of the day, regardless of whether one considers modern technology to be a blessing or a burden, the fact remains that it has the ability to improve cognitive capacities. The capacity to absorb new knowledge, to remember information, to employ logic, and to build connections between objects are all examples of these talents. They have an effect on both language and memory. The interactive educational games that can be downloaded onto mobile devices are something that we advocate. Some examples of these games include online puzzles, scribbling books, and other activities.

Restrictions imposed by electronic gadgets

The manufacturing and disposal of these goods have a considerable impact on the environment on a worldwide scale, despite the fact that a small electronic device at first glance may appear to be of little consequence. A wide variety of gadgets, if disposed of in an incorrect manner, may include potentially harmful elements including lead, cadmium, and mercury, which have the potential to pollute the soil surface. A further factor that makes the disposal problem worse is the pressure that is being put on to update to more modern equipment before it fails.

However, despite the fact that electronic gadgets have the ability to enhance communication, they are also held responsible for a number of societal problems. The proliferation of societal problems that are especially dangerous for children and teenagers is a direct result of the proliferation of internet-connected gadgets, including smartphones. Additionally, there is a growing concern that an undue reliance on smartphones and other electronic gadgets will result in an addiction to technology. Despite the fact that social media and other kinds of technology offer the ability to improve the connections of a large number of people, there are some people who feel forced to maintain a particular online identity or who experience melancholy as a result of their interactions on it.

GenZ on digital exposure

Among students, the addictions that manifest themselves most frequently are those that include texting, gaming, using the phone, or communicating via the internet. Not only do they stop acquiring new knowledge [4], but they also fail to complete their homework at home. In the past, students would spend their free time engaged in activities that were physically demanding, reading literature, or interacting with their classmates.

> **A lack of interest in academic pursuits**
>
> • Obesity in youngsters
>
> **Exposure to radiation for toddlers**
>
> • May induce violence in children

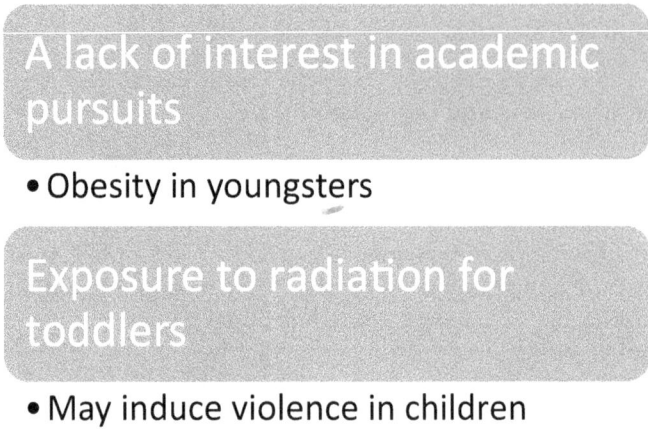

Figure 1.2 A lack of interest in academic pursuits.

However, engagement is gradually decreasing as a result of technological advancements, notably in the creative parts of participation. They have lost their sense of humor and their ability to be imaginative. The disadvantages of this exposure are listed in Figure 1.2 and are explained below.

At this point in time, it is up to the students themselves to decide whether they will develop poor study habits and a lazy attitude, or whether they will acquire information that is easily accessible and valuable. They commonly misspell terms since the majority of them use spell checkers, which causes them to misspell terms regularly. Furthermore, it leads to kids forgetting the stuff that they were first taught in school since they rely solely on the internet rather than being exposed to literature, and they seek aid from their lecturers.

Obesity

The condition known as obesity is the root cause of a wide range of serious health issues, including diabetes, heart attacks, and strokes. A significant portion of the calories consumed by children who spend an excessive amount of time in front of electronic gadgets is not completely burned off.

It's violent

It is possible for children to learn how to behave aggressively through the use of video games. Children who are overly involved in video games and other forms of technology are more likely to dispute or disobey their parents and teachers, according to research that has been conducted on the topic.

The exposure to radiations

There is a possibility that your infant will be exposed to radiofrequency. Indeed, you did hear that right. Due to the substantial quantity of radiation that modern appliances release, it is possible that continued use of these appliances in the house could be hazardous over time. It is of the utmost importance to limit the amount of time that children spend using technology and to safeguard them from radiation as this could be beneficial to their health.

There is no doubt that any item can be beneficial to your child if one is able to utilize it in a way that is both positive and instructional. Neither the implementation of such stringent rules nor the complete prohibition of technology is required or necessary. The maintenance of equilibrium is essential at every level. For the purpose of preventing ocular harm, it is recommended that young children be limited in the amount of time they spend in front of displays rather than being prohibited from using electronic devices altogether. One can find a plethora of internet activities that are not only instructive but also motivational.

WHAT IS RADIATION AND WHERE IS IT COMING FROM

Radiation is the term used to describe the process by which energy is transferred from one site to another in the form of pulses or particles. Radiation is something that many of us are exposed to on a daily basis. Some of the most well-known sources of radiation include the sun, microwave ovens in our kitchens, and the radios that we listen to while we are driving. The vast majority of this radiation does not pose any risk to human health. In general, radiation poses a lesser risk when administered at lower doses; however, when administered at larger doses, it can be associated with increased dangers. In order to protect our bodies and the environment from the harmful effects of radiation, we need to take a variety of precautions. At the same time, we need to be able to take advantage of the different applications of radiation, which vary depending on the kind of radiation.

What are the ways in which radiation might be beneficial? To give a few examples:

1. Radiation treatment is beneficial to our health since it enables us to undergo medical procedures, such as diagnostic imaging methods and a variety of cancer treatments.
2. Radiation makes it possible to generate electricity when used in conjunction with other energy sources, such as nuclear and solar energy.
3. Radiation can be used to create innovative plant kinds that are resistant to climate change or to clean wastewater. Both of these applications include the treatment of wastewater.

4. Nuclear techniques, which are based on radiation, provide scientists with the ability to investigate artifacts from the past or manufacture materials with superior properties, such as those used in the automobile industry. This is a significant contribution to both the scientific community and the industrial sector.

Given the potential benefits, why should we take precautions to protect ourselves from radiation?

Radiation has a wide range of uses that are beneficial; nonetheless, just like any other activity, it is essential to take special precautions to protect both the environment and human health whenever there are risks associated with the utilization of radiation. Various types of radiation require different precautions to be taken in order to protect oneself. In accordance with its mandate, the International Atomic Energy Agency (IAEA) establishes standards for the preservation of the environment and the people in relation to the peaceful use of ionizing radiation. For instance, a low-energy form of radiation [5], which is referred to as "non-ionizing radiation," may require fewer protective measures than a higher-energy form, which is referred to as "ionizing radiation."

Classes of radiation

Non-ionizing radiation

The following are examples of non-ionizing radiation: radio waves, microwaves, and visible light. The term "non-ionizing radiation" refers to radiation with a lower energy level that does not possess sufficient energy to separate electrons from atoms or molecules, regardless of whether the electrons are present in matter or in living creatures. Nevertheless, the energy that it possesses has the capability of causing the molecules to vibrate, which may result in the production of heat. Utilizing microwaves as an example, this is how microwave appliances function. The vast majority of people do not face any possible danger to their health from exposure to non-ionizing radiation. On the other hand, workers who are frequently subjected to particular kinds of non-ionizing radiation need to take additional precautions in order to protect themselves from the heat that is produced.

In addition to radio waves and visible light, there are other types of radiation that do not ionize molecules. Visible light is a type of non-ionizing radiation that may be perceived by the human eye. Radio waves are a type of non-ionizing radiation that cannot be perceived by the human eye or any of the other senses. However, traditional radios are able to intercept and decode radio waves.

Ionizing radiation

Some examples of ionizing radiation include radiation that is emitted from radioactive materials that are used in nuclear power plants and certain types of cancer therapies that involve gamma rays and X-rays. Ionizing radiation has the ability to detach electrons from atoms or molecules, causing changes to occur at the atomic level as a result of the interaction between matter, including living creatures, and the radiation. It is because such alterations often include the creation of ions, which are electrically charged atoms or molecules, that the term "ionizing" radiation was coined.

If we are exposed to large doses of ionizing radiation, it has the potential to bring about death or damage to the cells or organs in our bodies. When used at proper levels and with appropriate protective precautions, this type of radiation has a wide range of applications that are helpful. Some of these applications include the production of energy, in the industries and research, and the detection and treatment of various disorders, including cancer. The International Atomic Energy Agency (IAEA) provides a comprehensive system of international safety standards to assist legislators and regulators in protecting workers, patients, members of the public, and the environment from the potentially harmful effects of ionizing radiation. This is the case despite the fact that the national government is responsible for the regulation of the use of sources of radiation and radiation protection.

There is a clear correlation between the wavelength of ionizing and non-ionizing radiation and the amount of energy they possess. During the process of transitioning into a more stable state and releasing energy, for example, unstable (radioactive) atoms may produce ionizing radiation. It is because of the equilibrated and steady composition of particles (neutrons and protons) in their nucleus that the vast majority of atoms on Earth are stable. On the other hand, the fact that certain types of unstable atoms have a nucleus that is composed of a particular amount of protons and neutrons makes it impossible for them to keep the particles together. When radioactive atoms disintegrate, they emit energy in the form of ionizing radiation, which can be in the form of alpha particles, beta particles, gamma rays, or neutrons. These unstable atoms are typically referred to as "radioactive atoms." It is possible to harness and make use of this energy in a secure manner in order to accomplish a wide range of goals.

When referring to the process by which a radioactive atom becomes more stable through the release of particles and energy, the term "radioactive decay" is the term that is used to describe the process. To what extent does radioactive decay manifest itself in its most common forms? What preventative steps can we take to protect ourselves from the potentially harmful effects of the radiation that is produced as a secondary consequence?

Depending on the particles or waves that the nucleus emits in order to achieve stability, there are a variety of radioactive decay processes that can result in the emission of ionizing radiation. The four types of particles [6] that are most commonly found are neutrons, gamma rays, alpha particles, and beta particles.

Radiation of alpha

During the process of alpha radiation, the decaying nuclei release large particles that are positively charged in order to provide them with more stability. The penetration of these particles into our skin and the subsequent damage they cause can frequently be prevented by using merely a single sheet of paper. On the other hand, the inhalation of alpha-emitting substances by breathing, eating, or drinking might make interior tissues directly exposed, which can lead to adverse health effects. Across the globe, smoke detectors make use of americium-241, an atom that undergoes the process of alpha particle decay.

Radiation of the beta

During beta radiation, the nuclei emit electrons, which are smaller particles than alpha particles and have a greater ability to penetrate water layers. These electrons can travel through layers of water that are as thin as one to two centimeters, depending on the energy of the electrons. Aluminum pieces measuring a few millimeters in thickness are capable of successfully blocking beta radiation. Carbon-14 and hydrogen-3, often known as tritium, are two examples of unstable atoms that emit beta radiation. Tritium is used in emergency lamps for a variety of purposes, including illuminating exits in the dark and other purposes. This is because the beta radiation from tritium causes phosphor material to illuminate when it interacts with it, and this occurs without the use of electricity. Carbon-14, for example, is utilized in the process of determining the age of things that date back to the past.

Radiation of the gamma

Because of its ability to produce gamma radiation, which may be used to eliminate tumors, cobalt-60 is utilized in the diagnosis and treatment of cancer. Beginning with the emission of beta radiation, the nucleus of cobalt-60 transforms into nickel-60 in a highly energetic state, which is denoted by the symbol *Ni-60. In this state, energy is rapidly dissipated through the emission of two gamma rays, which ultimately results in nickel-60 that is stable. Gamma rays, which are classified as electromagnetic radiation, have a wide range of applications, one of which is the treatment of cancer. It is comparable to X-rays in this regard. The human body is capable of absorbing certain gamma rays, which can lead to injuries, whereas other gamma rays can pass through the human body without causing any harm. The intensity of gamma radiation can be reduced to a level that is less dangerous by using materials such as concrete or lead to construct thick walls. Radiotherapy treatment rooms in hospitals and other facilities that provide care for cancer patients have walls that are significantly thicker than average because of this reason.

Neutrons

Within a nuclear reactor, nuclear fission presents an example of a radioactive chain reaction that is maintained by neutrons. Neutrons are particles that are rather large and are one of the fundamental components of the nucleus. Because they are not charged, they do not immediately create ionization. However, the interaction of these particles with the atoms of matter might result in the production of alpha, beta, gamma, or X-rays, which then lead to the ionization of the matter. It is only through the use of substantial amounts of concrete, water, or paraffin that it is possible to stop the penetration of neutrons. Nuclear reactors and nuclear reactions that are triggered by high-energy particles in accelerator beams are two examples of the many ways that neutrons can be produced. Neutrons can also be produced through a variety of other processes. Nucleons have the potential to be a significant source of radiation that is indirectly ionized.

The role of International Atomic Energy Agency (IAEA)

The International Atomic Energy Agency (IAEA) [7] offers assistance to its member states in the implementation of nuclear technology, such as radiation, for the purposes of health, agriculture, environmental protection, water management, energy, and industry. The IAEA is responsible for coordinating research efforts and implementing projects in countries all over the world. Additionally, it provides assistance in the research and development of utilizing radiation and radioactive sources in practical applications. By implementing safeguards and conducting verification activities, the IAEA ensures that materials capable of emitting radiation are not diverted from peaceful purposes. It is also responsible for establishing safety standards and security recommendations, as well as providing reports on the most effective techniques for protecting individuals, society, and the environment from the harmful effects of ionizing radiation.

POWER CONSTRAINTS AND NECESSITIES IN DOMESTIC AND COMMERCIAL LIVES

The pervasive presence of wireless devices in our lives has revolutionized how we communicate, work, and manage our daily activities. These devices, from smartphones and smartwatches to home automation systems and industrial IoT solutions, rely heavily on a continuous power supply to function effectively. This dependence on power brings forth significant constraints and necessities, emphasizing the importance of efficient power management and addressing the associated electromagnetic exposures.

In domestic settings, the convenience offered by wireless devices comes at the cost of constant power needs. Smart home systems, including lighting, security cameras, and thermostats, require uninterrupted power to operate seamlessly. For instance, a smart thermostat like the Nest Learning Thermostat not only monitors and adjusts the home temperature but also connects to the internet to provide remote control and energy usage statistics. Without a reliable power supply, such devices lose functionality, rendering them ineffective and defeating the purpose of a connected home.

Power constraints are also evident in the use of wearable devices. Fitness trackers and smartwatches, such as the Apple Watch, require regular charging to maintain their health monitoring and communication features. The need to frequently recharge these devices underscores the challenge of battery life and power efficiency. Advanced wireless charging solutions and battery technologies are continually being developed to mitigate these constraints, but the issue of power dependence remains.

In commercial environments, the reliance on wireless devices is even more pronounced. Businesses utilize a wide range of wireless technologies to streamline operations, enhance productivity, and improve customer service. For instance, retail stores employ wireless point-of-sale (POS) systems to facilitate quick and efficient transactions. These systems rely on a stable power supply to process payments, manage inventory, and perform other critical functions. Any disruption in power can lead to significant operational setbacks and financial losses.

Similarly, industrial sectors leverage wireless IoT devices for monitoring and controlling machinery, tracking assets, and ensuring safety. These devices are integral to the efficiency and productivity of industrial operations. However, they also necessitate a robust power infrastructure to maintain continuous operation. The use of energy-efficient power supplies and backup power solutions is crucial to mitigate the risk of power failures and ensure the reliability of these systems.

The widespread use of wireless devices and their power requirements also raise concerns about electromagnetic exposure. Continuous operation of wireless devices generates electromagnetic fields (EMFs), which have been the subject of extensive research regarding their potential health impacts. According to the World Health Organization (WHO), while the evidence does not confirm any health consequences from low-level exposure to EMFs, it is essential to adhere to established safety guidelines to mitigate any potential risks. For instance, maintaining a safe distance from high-power devices and using shielded cables can help reduce exposure.

Moreover, the Environmental Protection Agency (EPA) [8] recommends minimizing unnecessary exposure by turning off wireless devices when not in use and using wired connections where possible. These measures are particularly important in environments with high densities of wireless devices, such as offices and schools, where cumulative exposure can be significant.

HUMAN NERVOUS SYSTEM

Extensive research has been conducted over the past century to understand the impact of ionizing radiation on human physiology, particularly in the aftermath of nuclear accidents and incidents involving the absorption or ingestion of radioactive material [9]. Artificial sources of ionizing radiation are increasingly valuable alongside exposure to the environment. A significant portion of Australia's annual radiation exposure is now attributed to ionizing radiation from medical diagnostic and treatment procedures [10]. A reassessment of the literature on the biological effects of ionizing radiation at different doses is required because of the increasing accessibility and usefulness of ionizing radiation in medical applications, particularly in relation to the central nervous system (CNS). Extensive research has been conducted on the impact of high-dose ionizing radiation exposure on the nervous system. However, there has been considerable debate surrounding the effects of low-dose exposure [11]. It is widely acknowledged that biological systems exhibit a linear dose response when exposed to ionizing radiation. The existing animal radiobiological data pose a challenge to the widely accepted linear no-threshold model. Multiple lines of evidence suggest that exposure to lower levels of ionizing radiation may provide neuroprotection [12]. This reaction has been described as radiation hormesis, where there is a debate about whether minimal exposure to a stressor can trigger radioadaptive, reparative, and protective processes. The concept of radiation hormesis is not currently acknowledged by international panels and regulating bodies. This is primarily because there is insufficient evidence to support the idea that low-dose ionizing radiation has beneficial effects on human physiology. Furthermore, there is a lack of consensus on what exactly qualifies as a "low dose." While the biological responses to high and low doses of ionizing radiation remain unclear, ongoing research is shedding light on the molecular and cellular mechanisms involved in radiobiological reactions, particularly in relation to the complex and multifaceted central nervous system.

The field of radiation science has greatly contributed to advancements in medical practices since its initial discovery. However, it wasn't until much later in history that researchers began directly examining the effects of ionizing radiation on the brain. Using ionizing radiation for medical purposes can be a bit contradictory since higher doses can potentially harm healthy tissue. This analysis will focus on the non-cancerous effects of ionizing radiation exposure and the cellular responses it can induce in the adult central nervous system. The hippocampus [13], a region known for its sensitivity to radiation, has been extensively studied for its potential harm caused by exposure. When differentiated cells become part of the hippocampus network, excessive radiation can lead to cell malfunction or apoptosis. This, in turn, can cause long-term functional issues. Microglial-mediated neuroinflammation and oxidative stress caused by an overabundance of reactive oxygen species (ROS) production coordinate responses to high-dose irradiation [14]. Microglial responses and the balance of mitochondrial redox are essential factors in controlling the body's response to low-dose radiation. They primarily work by enhancing antioxidant defenses. There is a wealth of evidence indicating that using lower doses can have a positive impact on cell function, molecular structures, synapses, and important brain mechanisms like neurogenesis. It has also been found that lower doses can trigger reparative mechanisms in the presence of CNS pathology, although there is still some evidence suggesting a linear dose-response pattern. An acute exposure to less than 100mSv, or 0.1 Gy, is considered a low dose, according to recommendations from regulatory bodies and findings from research programs focused on low-dose radiation.

COMPATIBILITY OF HUMAN BODY AND WIRELESS COMMUNICATION

The integration of the human body with wireless communication technology is becoming increasingly pivotal in modern healthcare. With the rise in chronic diseases and an aging population, there is a growing need for continuous health monitoring to effectively manage and enhance the quality of life for patients. Body Area Networks (BANs) are at the forefront of this innovation, providing real-time monitoring of physiological data and offering therapeutic functionalities. BANs consist of miniature sensors and actuators placed on or within the body to collect vital data. This data is then wirelessly transmitted to a relay node or aggregator, such as a smartwatch or smart wristband, and subsequently forwarded to a central control system for analysis and action.

Traditional wireless communication methods, including Zigbee, Bluetooth, and Ultrawide Band, face significant challenges in human-centric applications. These techniques are susceptible to electromagnetic interference, signal leakage, and eavesdropping, and they are impeded by the body's high water content, which blocks radio frequency (RF) signals. Inductive wireless coupling, while an alternative, suffers from low coupling efficiency and requires larger coil sizes, making it less practical for compact and wearable devices.

Human Body Communication (HBC) [15] emerges as a groundbreaking transmission technique that utilizes the human body itself as a medium for electrical signal transmission. HBC boasts several advantageous characteristics, including minimal signal leakage, high security, low signal attenuation, and low carrier frequency. These features enable lower transmission power, reduced power consumption, and the potential for device miniaturization. Consequently, HBC is well-suited for BANs, offering low interference, high security, appropriate transmission range, and the potential for smaller, more efficient devices. To optimize HBC for healthcare applications, it is essential to minimize transmission power and enhance communication performance while ensuring reliable data transmission. Understanding the channel characteristics and communication dynamics of HBC is crucial for its effective deployment in healthcare settings.

In today's world, the human race has become increasingly dependent on gadgets for daily functioning and well-being. This dependence extends to healthcare, where wearable devices such as fitness trackers, smartwatches, and health monitors have become indispensable. These gadgets continuously collect data on various health parameters, such as heart rate, blood pressure, glucose levels, and physical activity. For instance, a diabetic patient can use a continuous glucose monitor (CGM) that wirelessly transmits glucose readings to a smartphone app, enabling real-time monitoring and timely insulin adjustments. Similarly, smartwatches equipped with electrocardiogram (ECG) sensors can detect irregular heart rhythms and alert the wearer to seek medical attention, potentially saving lives.

Beyond healthcare, gadgets have permeated every aspect of daily life. Smartphones serve as personal assistants, navigation aids, entertainment hubs, and communication tools. Home automation devices like smart thermostats, lights, and security systems provide convenience and enhanced control over living environments. Virtual assistants, such as Amazon's Alexa and Google Assistant, streamline tasks and information retrieval, making daily routines more efficient.

This pervasive reliance on gadgets underscores the necessity of seamless and reliable wireless communication technologies. As healthcare continues to evolve, integrating advanced communication techniques like HBC into BANs can further enhance the effectiveness and security of health monitoring systems. The future of healthcare lies in leveraging these innovative technologies to create a more connected, responsive, and personalized approach to patient care.

REFERENCES

1. Sarla, G. S. (2019). Excessive use of electronic gadgets: health effects. *The Egyptian Journal of Internal Medicine*, *31*(4), 408–411. https://doi.org/10.4103/ejim.ejim_56_19
2. Jadad, A. R. (2003). From electronic gadgets to better health: where is the knowledge? *BMJ*, *327*(7410), 300–301. https://doi.org/10.1136/bmj.327.7410.300

3. Garg, S., Aggarwal, D., Upadhyay, S. K., Kumar, G., & Singh, G. (2020). Effect of covid-19 on school education system: challenges and opportunities to adopt online teaching and learning. *Humanities & Social Sciences Reviews*, *8*(6), 10–17. https://doi.org/10.18510/hssr.2020.862

4. Djafarova, E., & Foots, S. (2022). Exploring ethical consumption of generation Z: theory of planned behaviour. *Young Consumers Insight and Ideas for Responsible Marketers*, *23*(3), 413–431. https://doi.org/10.1108/yc-10-2021-1405

5. Vesley, D. (1999). Ionizing and nonionizing radiation. In *Human health and the environment* (pp. 65–74). Springer. https://doi.org/10.1007/978-1-4757-5434-6_7

6. Choppin, G., Liljenzin, J. O., & Rydberg, J. (2002). *Radiochemistry and nuclear chemistry*. Butterworth-Heinemann.

7. Fischer, D. (1997). *History of the international atomic energy agency*. United Nations Publications.

8. National Research Council, Policy and Global Affairs, Science and Technology for Sustainability Program, & Committee on Incorporating Sustainability in the U.S. Environmental Protection Agency. (2011). *Sustainability and the U.S. EPA*. National Academies Press.

9. Aub, J. C., Evans, R. D., Hempelmann, L. H., & Martland, H. S. (1952). The late effects of internally-deposited radioactive materials in man. *Medicine*, *31*(3), 221–329. https://doi.org/10.1097/00005792-195209000-00001

10. Gerber, T. C., Carr, J. J., Arai, A. E., Dixon, R. L., Ferrari, V. A., Gomes, A. S., Heller, G. V., McCollough, C. H., McNitt-Gray, M. F., Mettler, F. A., Mieres, J. H., Morin, R. L., & Yester, M. V. (2009). Ionizing radiation in cardiac imaging. *Circulation*, *119*(7), 1056–1065. https://doi.org/10.1161/circulationaha.108.191650

11. Needleman, H. L., Schell, A., Bellinger, D., Leviton, A., & Allred, E. N. (1990). The long-term effects of exposure to low doses of lead in childhood. *The New England Journal of Medicine*, *322*(2), 83–88. https://doi.org/10.1056/nejm199001113220203

12. Xiong, Z. G., Zhu, X. M., Chu, X. P., Minami, M., Hey, J., Wei, W. L., MacDonald, J. F., Wemmie, J. A., Price, M. P., Welsh, M. J., & Simon, R. P. (2004). Neuroprotection in ischemia. *Cell*, *118*(6), 687–698. https://doi.org/10.1016/j.cell.2004.08.026

13. Eriksson, P. S., Perfilieva, E., Björk-Eriksson, T., Alborn, A. M., Nordborg, C., Peterson, D. A., & Gage, F. H. (1998). Neurogenesis in the adult human hippocampus. *Nature Medicine*, *4*(11), 1313–1317. https://doi.org/10.1038/3305

14. Yukihara, E., Whitley, V., McKeever, S., Akselrod, A., & Akselrod. (2004). Effect of high-dose irradiation on the optically stimulated luminescence of Al2O3:C. *Radiation Measurements*, *38*(3), 317–330. https://doi.org/10.1016/j.radmeas.2004.01.033

15. Zhao, J. F., Chen, X. M., Liang, B. D., & Chen, Q. X. (2016). A review on human body communication: signal propagation model, communication performance, and experimental issues. *Wireless Communications and Mobile Computing*, *2017*(1), 5842310. https://doi.org/10.1155/2017/5842310

Chapter 2

Electromagnetic basics in terms of E and H energy

ELECTROMAGNETIC BASICS IN TERMS OF E AND H ENERGY

Electromagnetism, the force that governs everything from light and radio waves to the very building blocks of matter, is a fascinating interplay between two fundamental fields: the electric field (E) and the magnetic field (H). These fields, though seemingly separate, are intricately linked, carrying and transferring energy throughout the universe.

Electric fields: the push and pull of charge

Imagine a charged object, like a positively charged ball. It exerts an invisible force on its surroundings, pushing away other positive charges and attracting negative ones. This invisible force is the electric field (E), described by equation 2.1:

$$E = F/q \tag{2.1}$$

where E is the electric field strength (measured in volts per meter, V/m), F is the electric force experienced by a test charge (measured in Newtons, N), and q is the magnitude of the test charge (measured in Coulombs, C). The strength of the E field depends on the object's charge and weakens with distance following an inverse square law as stated in equation 2.2:

$$E \propto 1/r^2 \tag{2.2}$$

where r is the distance from the charged object.

Electric fields have the ability to store energy. The energy density (u_e) associated with an electric field is given by equation 2.3:

$$u\_e = 1/2 * \varepsilon * E^2 \tag{2.3}$$

DOI: 10.1201/9781003590712-2

where ε (epsilon) is the permittivity of the medium (measured in Farads per meter, F/m). This equation shows that the energy density increases with the square of the electric field strength.

Magnetic fields: a whirlwind of moving charges

Now, imagine a current (I) flowing through a wire. This movement of charged particles, like electrons, creates a swirling magnetic field (H) around the wire. The relationship between current and magnetic field is described by Ampere's Law as stated in equation 2.4:

$$\oint H \cdot dl = I \tag{2.4}$$

where $\oint$ (integral sign) represents a closed loop around which the line integral is taken, dl is a small element of length along the loop, and $H \cdot dl$ is the dot product of the magnetic field strength (H) and the length element. This equation states that the circulation of the magnetic field around a closed loop is equal to the current enclosed by the loop. The magnetic field, unlike the electric field, cannot exist alone. However, it can indirectly store energy. This stored energy is not within the H field itself, but rather in the motion of the charges that generate it. For example, a strong current in a coil of wire creates a strong H field, which indirectly represents the kinetic energy of the moving charges.

The E and H tango: a beautiful partnership

The magic of electromagnetism lies in the relationship between E and H fields. A changing electric field can induce a magnetic field, and vice versa. This relationship is described by Faraday's law of induction as seen in equation 2.5:

$$\nabla \times E = -\partial B / \partial t \tag{2.5}$$

where $\nabla \times$ (nabla cross) is a mathematical operator that calculates the curl of a vector field, E is the electric field, B is the magnetic field (related to H by $B = \mu * H$, where μ is the permeability of the medium), and $\partial B / \partial t$ represents the rate of change of the magnetic field with time. This equation states that a changing electric field induces a curl in the magnetic field. Together, Faraday's Law and Ampere's Law form the foundation of Maxwell's equations, which govern the behavior of electromagnetic fields. These changing fields create a self-propagating wave, where a change in one field creates a change in the other, and so on. These waves, known as electromagnetic waves, carry energy through space at the speed of light (c), according to equation 2.6:

$$c = 1/\sqrt{(\varepsilon \, \mu)} \tag{2.6}$$

where ε and μ are the permittivity and permeability of the medium, respectively. From radio waves to X-rays, all forms of light are electromagnetic waves with different frequencies and energies.

Energy transfer with the Poynting vector

The flow of energy in an electromagnetic field is described by the Poynting vector (S). This vector is given by equation 2.7:

$$S = E \times H \tag{2.7}$$

where S is the Poynting vector (measured in watts per square meter, W/m^2), E is the electric field, and H is the magnetic field. The Poynting vector points in the direction of energy propagation and its magnitude represents the power (energy per unit time) carried by the wave.

Electromagnetic applications: energy at work

The concept of E and H field energy plays a crucial role in various applications. Radio antennas use changing E and H fields to transmit and receive electromagnetic waves, carrying information across vast distances. Power lines transfer electrical energy using strong E and H fields, while transformers manipulate these fields to adjust voltage levels. Even electric motors rely on the interaction between E and H fields to produce torque and rotation.

Beyond the basics: a universe of possibilities

Understanding the relationship between E and H fields is just the first step in exploring the vast realm of electromagnetism. More advanced concepts like Maxwell's equations and electromagnetic induction delve deeper into this fascinating phenomenon. These equations predict the behavior of E and H fields under various circumstances and explain how they interact with matter to create a wide range of phenomena, from light refraction to the operation of electronic devices.

APPLICATIONS AND EFFECTIVENESS OF RADIATING POWER

Radiating power, the transfer of energy through electromagnetic waves, has revolutionized numerous aspects of our lives. From wireless communication to medical imaging, this technology offers unique advantages and challenges. Radiating power relies on the fundamental principles of electromagnetism. Maxwell's equations, a set of four partial differential equations [1], govern the behavior of electric and magnetic fields (E and H):

- Gauss's law for electricity: $\nabla \cdot E = \rho/\varepsilon_0$ (relates electric field divergence to charge density)
- Gauss's law for magnetism: $\nabla \cdot B = 0$ (magnetic monopoles don't exist)
- Faraday's law of induction: $\nabla \times E = -\partial B/\partial t$ (changing magnetic field induces electric field)
- Ampere's law with Maxwell's addition: $\nabla \times H = J + \partial E/\partial t$ (current density and changing electric field create magnetic field)

In these equations, ∇ is the divergence operator, $\nabla \times$ is the curl operator, ρ is the charge density, ε_0 is the permittivity of free space, B is the magnetic field (related to H by $B = \mu_0 H$, where μ_0 is the permeability of free space), J is the current density, and $\partial/\partial t$ represents the partial derivative with respect to time.

These equations predict the behavior of electromagnetic waves, characterized by their wavelength (λ), frequency (f), and speed (c). The relationship between these is expressed in [2] by equations 2.8 and 2.9:

$$c = \lambda f \text{ (where } c \approx 3 \times 10^8 \text{ m/s)} \tag{2.8}$$

The power (P) carried by an electromagnetic wave is related to its electric and magnetic field strengths through the Poynting vector (S) as in equation 2.9:

$$S = (1/\mu_0) * E \times H \text{ (where S is in W/m}^2) \tag{2.9}$$

The Poynting vector points in the direction of energy propagation, and its magnitude represents the power density of the wave.

Applications of radiating power

Radiating power finds applications in diverse fields, each with specific requirements for wavelength, power density, and efficiency. Here are some key examples:

- *Wireless communication*: Radio waves (wavelength, $\lambda \approx$ mm to km) are the workhorse of wireless communication, enabling cellular networks, Wi-Fi, Bluetooth, and satellite communication. Different frequency bands cater to varying ranges, data rates, and penetration capabilities. For example, lower frequencies penetrate buildings better but offer lower data rates compared to higher frequencies [3].

- *Radio frequency (RF) heating*: RF waves ($\lambda \approx$ mm to cm) can penetrate materials and heat them internally due to dielectric heating or magnetic induction. This principle finds applications in industrial drying, food processing, and medical hyperthermia treatments for tumors [4].
- *Microwave ovens*: Microwaves ($\lambda \approx$ mm to cm) are used for rapid and efficient cooking. They interact with water molecules in food, causing them to vibrate and generate heat. The specific frequency chosen ensures optimal penetration depth and heating uniformity within the food [5].
- *Radar and remote sensing*: Radar systems ($\lambda \approx$ mm to m) use radio waves to detect and track objects. The reflected waves provide information about the target's distance, speed, and size. Similarly, remote sensing applications utilize various wavelengths to study Earth's surface, monitor weather patterns, and map resources [6].
- *Electromagnetic spectrum imaging*: Different portions of the electromagnetic spectrum offer unique insights into materials and biological structures. X-rays ($\lambda \approx$ pm to nm) are used in medical imaging for bone fractures and tumors. Infrared radiation ($\lambda \approx$ μm to mm) finds applications in night vision and thermal imaging [7].

Effectiveness and considerations

The effectiveness of radiating power depends on several factors:

- *Wavelength*: The chosen wavelength determines the interaction with the target material. Shorter wavelengths like X-rays have higher penetration capabilities but can be harmful to biological tissues. Longer wavelengths like microwaves penetrate deeper but may be less effective for specific applications.
- *Power density*: The intensity of the radiation, defined by the Poynting vector, determines the rate of energy transfer. Higher power densities can achieve faster heating or stronger signals but require careful consideration to avoid safety hazards.
- *Transmission efficiency*: The efficiency of radiating power transfer depends on factors like distance, intervening obstacles, and the receiving antenna's characteristics. Signal attenuation with distance can be a challenge, especially for longer wavelengths. Additionally, objects in the path can absorb or scatter the radiation, reducing its effectiveness.
- *Safety considerations*: Exposure to high-power radiation, particularly ionizing radiation like X-rays, can pose health risks. Regulations and safety standards govern exposure limits in various applications to minimize these risks. For example, medical X-ray procedures use collimators to focus the radiation on the target area, reducing exposure to surrounding tissues.

ENERGY HARVESTING

Energy harvesting (EH) is a rapidly developing technology that captures and converts ambient energy from the environment into usable electrical power. This approach offers a sustainable and potentially limitless power source for low-power electronic devices, reducing dependence on batteries and wired connections.

Core principles and energy sources

Energy harvesting relies on the principle of converting various forms of ambient energy into electricity. The energy sources include:

Solar energy: Photovoltaic (PV) cells convert sunlight into electrical power using the photovoltaic effect.

Thermal energy: Thermoelectric generators (TEGs) exploit the Seebeck effect, generating voltage due to a temperature difference between two junctions.

Vibrational energy: Piezoelectric materials produce a voltage when mechanically stressed due to the piezoelectric effect.

Kinetic energy: Electromagnetic induction allows electromagnetic generators (EMGs) to convert mechanical motion into electricity.

Radio frequency (RF) energy: Ambient RF waves can be harvested using rectenna antennas, converting them into DC power.

The choice of harvesting technique depends on the available energy source and the power requirements of the device. Different techniques offer varying power densities, making them suitable for specific applications.

Energy harvesting techniques

1. *Solar energy harvesting*

Solar energy is a widely utilized source for EH. Photovoltaic (PV) cells, made from materials like silicon, convert sunlight into electrical power. The efficiency of PV cells determines the amount of electricity generated, with advanced cells achieving efficiencies exceeding 20% [8]. Solar energy harvesting finds applications in powering sensors, wearables, and small electronics in outdoor environments.

2. *Thermal energy harvesting*

Thermal energy harvesting utilizes thermoelectric generators (TEGs) that convert heat flow into electricity. TEGs exploit the Seebeck effect, where a temperature difference between two junctions creates a voltage difference. Their efficiency depends on the materials used and the temperature gradient. TEGs are suitable for applications with a consistent heat source, such as waste heat from industrial processes or body heat from wearables.

3. *Vibrational energy harvesting*

Piezoelectric materials can convert mechanical stress into electrical energy due to the piezoelectric effect. Vibrations from sources like machinery, traffic, or human motion can be harvested using piezoelectric transducers [9]. This technique finds applications in powering wireless sensor nodes or self-powered structural health monitoring systems.

4. *Kinetic energy harvesting*

Kinetic energy harvesting utilizes electromagnetic generators (EMGs) to convert mechanical motion into electricity. These generators rely on Faraday's law of induction, where a changing magnetic field within a coil induces a voltage. EMGs can be used to harvest energy from human motion, wind, or water flow. Applications include powering self-powered wireless keyboards or powering environmental monitoring sensors.

5. *Radio frequency energy harvesting*

Ambient radio frequency (RF) energy from radio and television broadcasts, Wi-Fi signals, or cellular networks can be harvested using rectenna antennas. These antennas are designed to capture and rectify (convert) the RF waves into usable DC power. The power density of RF energy is typically low, making it suitable for powering low-power devices like wearables or remote sensors in environments with strong RF signals.

Applications of energy harvesting

Energy harvesting [10, 11] offers a promising solution for powering various low-power electronic devices across diverse fields. Here are some key application areas:

Wireless sensor networks (WSNs): Energy harvesting can eliminate battery replacements in sensor nodes deployed for environmental monitoring, structural health monitoring, and industrial automation. This reduces maintenance costs and ensures long-term functionality.

Wearable electronics: Energy harvesting from solar, thermal, or kinetic sources can power wearable devices like smartwatches, fitness trackers, and health monitors. This enables continuous operation and eliminates the need for frequent charging [12].

Internet of Things (IoT): Energy harvesting can power low-power IoT devices used for smart homes, asset tracking, and industrial automation. This facilitates the creation of self-powered and interconnected networks [13].

Implantable medical devices: Energy harvesting from body heat or motion can power implantable medical devices like pacemakers and glucose monitors. This eliminates the need for invasive battery replacements, improving patient comfort and safety.

Remote sensing: Energy harvesting can power remote sensors deployed in areas with limited access to grid power, such as for environmental monitoring in remote locations or agricultural monitoring in fields [6].

Looking ahead: challenges and future directions

While energy harvesting offers significant benefits, there are still challenges to address. These include:

Low power density: Ambient energy sources often have lower power density compared to conventional sources like batteries. This necessitates efficient energy management techniques and careful device design for low-power operation.

Intermittency: Some energy sources, like solar or wind, are intermittent. Energy storage solutions like capacitors or small batteries are crucial to bridge periods of low energy availability.

Cost and efficiency: Developing highly efficient energy harvesting technologies can be expensive. Continued research and development are needed to bring down costs and improve efficiency for wider adoption.

Despite these challenges, energy harvesting holds immense potential for powering the future of low-power electronics. As technology advances and costs decrease, we can expect to see a wider range of self-powered devices that contribute to a more sustainable and interconnected world.

GREEN ENERGY SOLUTIONS

The world faces a critical challenge: meeting growing energy demands while mitigating the environmental impacts of traditional fossil fuel-based energy sources. Green energy solutions, also known as renewable energy sources, offer a promising path toward a sustainable future. These solutions tap into naturally replenished resources like sunlight, wind, water, biomass, and geothermal heat to generate electricity and power various applications.

Solar energy

Solar energy, harnessed from the sun's rays, is one of the most abundant and rapidly growing renewable energy sources. It can be captured using two primary technologies.

Photovoltaic (PV) cells: These cells convert sunlight directly into electricity using the photovoltaic effect. Silicon is the most common material used in solar panels, but research is ongoing to develop more efficient and cost-effective materials.

Concentrating solar power (CSP): CSP systems use mirrors to concentrate a large area of sunlight onto a small area. This concentrated light is then used as heat to drive a traditional power plant.

Technical aspects

PV systems come in various sizes, from rooftop panels for individual homes to large-scale solar farms generating electricity for entire communities. The efficiency of PV panels determines the amount of electricity generated per unit of sunlight received. Efficiency varies depending on the material and technology used, with current commercially available panels reaching efficiencies exceeding 20%. Factors like location, sunlight availability, and panel orientation impact the overall energy production of a solar system.

Real-life examples

Rooftop solar panels: Millions of homeowners and businesses worldwide have installed rooftop solar panels to generate electricity for their own use. This reduces reliance on the grid and lowers electricity bills.

Utility-scale solar farms: Large-scale solar farms can provide clean energy for entire towns and cities. For instance, the Solar Star Project in California is one of the world's largest operating photovoltaic power plants, generating enough power for over 250,000 homes.

Benefits

- Abundant and readily available solar energy source.
- Reduces reliance on fossil fuels and greenhouse gas emissions.
- Scalable solutions for individual and large-scale power generation.
- Technological advancements are continuously improving efficiency and reducing costs.

Challenges

- Intermittent energy source—sunlight is not available 24/7, requiring energy storage solutions for grid stability.
- Initial installation costs can be high, although they have significantly decreased in recent years.

- Solar panel efficiency needs further improvement to maximize energy production.

Wind energy

Wind energy utilizes the power of moving air to generate electricity. Wind turbines convert the kinetic energy of wind into electricity using rotating blades connected to a shaft and generator.

Technical aspects

Wind turbines come in various sizes, from small turbines for homes to large turbines used in wind farms. The size and location of the turbine significantly impact the amount of electricity generated. Stronger and more consistent winds lead to higher energy production.

Wind farms are often located in open areas with good wind resources, such as coastal regions or high plains.

Real-life examples

Offshore wind farms: Offshore wind farms positioned in ocean waters with stronger and steadier winds are becoming increasingly popular. For example, the Hornsea One wind farm in the UK is one of the world's largest offshore wind farms, generating enough electricity to power over 1 million homes.

Distributed wind power: Smaller wind turbines can be installed in rural or agricultural areas to provide electricity for individual homes or communities.

Benefits

- Abundant and renewable wind energy source.
- Reduces reliance on fossil fuels and greenhouse gas emissions.
- Technological advancements are leading to larger and more efficient wind turbines.
- Wind farms can create jobs in construction, operation, and maintenance.

Challenges

- Wind availability can be variable, impacting energy production.
- Wind farms may have visual and noise impacts on nearby communities.
- Concerns exist regarding the impact of wind turbines on bird migration patterns.

Hydropower

Hydropower harnesses the energy of moving water to generate electricity. Hydroelectric dams utilize falling water to drive turbines connected to generators.

Technical aspects

Hydropower plants can be classified as run-of-the-river, pumped storage, or dam-based depending on their design and operation. Run-of-the-river plants divert a portion of a river's flow through turbines without creating a large reservoir. Pumped storage plants use electricity during off-peak hours to pump water uphill into a reservoir. This stored water can then be released through turbines to generate electricity during peak demand periods.

Dam-based hydropower plants create large reservoirs to store water and generate electricity on demand.

Real-life examples

Three Gorges Dam: Situated in China, the Three Gorges Dam is the world's largest hydroelectric facility in terms of installed capacity. It generates clean electricity for millions of homes and businesses.

Small hydropower projects: Small-scale hydropower installations can provide electricity for remote communities or specific applications like powering irrigation systems.

Benefits

- Mature and reliable renewable energy source with predictable power generation.
- Hydropower dams can also provide flood control and water storage benefits.
- Pumped storage offers energy storage capabilities to support grid stability with intermittent renewable sources.

Challenges

- Construction of large dams can have significant environmental impacts on ecosystems and river flows.
- Resettlement of communities displaced by dam construction can be a social concern.
- Run-of-the-river and tidal hydropower projects may have lower energy production capacities compared to large dams.

Geothermal energy

Geothermal energy utilizes the Earth's internal heat to generate electricity. This heat source can be accessed through geothermal power plants that use steam or hot water from underground reservoirs to drive turbines connected to generators.

Technical aspects

Geothermal resources vary depending on location. High-temperature resources are ideal for electricity generation, while lower-temperature resources can be used for direct heating applications. Geothermal power plants are typically located in areas with geological features like volcanic activity or hot springs, indicating geothermal potential.

Real-life examples

Geysers Geothermal Field: Located in California, the Geysers Geothermal Field is the world's largest geothermal complex for electricity generation. It has been in operation since the 1960s and continues to provide clean energy to the region.

Direct geothermal heating: Geothermal energy can be used for direct heating applications in buildings, greenhouses, and district heating systems. This reduces reliance on conventional heating methods.

Benefits

- Renewable and reliable energy source with minimal emissions.
- Geothermal power plants have a smaller land footprint compared to some renewable energy sources.
- Geothermal heating offers a clean and efficient alternative to traditional heating systems.

Challenges

- Usable geothermal resources are geographically limited and require exploration to identify suitable locations.
- Initial development costs for geothermal power plants can be high.
- Improper well management can lead to environmental concerns like ground subsidence or water contamination.

Biomass energy

Biomass energy utilizes organic matter from plants, animals, and waste materials as a fuel source. This biomass can be converted into various forms of energy, including electricity, heat, and transportation fuels.

Technical aspects

Biomass can be used directly for combustion in power plants or converted into biofuels like biodiesel or biogas through processes like fermentation. Different types of biomass resources exist, including dedicated energy crops, agricultural residues, and municipal solid waste.

Real-life examples

Biogas power plants: Biogas plants capture methane produced from the decomposition of organic waste in landfills or agricultural operations. This biogas can then be used to generate electricity or heat.

Biofuel production: Biofuels derived from biomass resources like corn or sugarcane can be used as a cleaner alternative to traditional fossil fuels for transportation.

Benefits

- Renewable energy sources can utilize waste materials and reduce reliance on landfills.
- Biofuels can offer a lower carbon footprint compared to traditional gasoline or diesel.

Challenges

- Sustainable biomass production practices are crucial to avoid deforestation and land-use change.
- Efficiency of biomass conversion processes needs improvement to maximize energy output.
- Concerns exist regarding competition between food production and biomass feedstock crops.

Ocean energy

The power of the ocean can be harnessed through various technologies to generate clean energy. These technologies include:

Tidal energy: Utilizing the rise and fall of tides to generate electricity using underwater turbines.

Wave energy: Capturing the energy of waves using devices that convert wave motion into electricity.

Ocean thermal energy conversion (OTEC): Exploiting the temperature difference between deep and surface ocean waters to generate electricity using a closed-loop system.

Technical aspects

Ocean energy technologies are still under development, and their commercial viability is being explored. Different ocean energy technologies are suited for specific oceanographic conditions like tidal currents or wave patterns.

Real-life examples

Sihanet Tidal Farm: Located in South Korea, the Sihanet Tidal Farm is one of the world's largest operating tidal energy projects. It utilizes underwater turbines to generate electricity from tidal currents.

Wave energy pilot projects: Several pilot projects worldwide are testing and demonstrating various wave energy conversion technologies to assess their feasibility and efficiency.

Benefits

- Renewable and predictable energy source with high potential, especially in coastal regions.
- Ocean energy can contribute to diversifying the energy mix and reducing reliance on traditional sources.
- Technological advancements have the potential to make ocean energy a cost-competitive renewable option.

Challenges

- Ocean energy technologies are still in their early stages of development and require further research and innovation.
- The harsh marine environment poses challenges for the durability and maintenance of ocean energy devices.
- Environmental impacts on marine life and ecosystems need careful consideration during project development.

Hence, considering all of the above, our understanding of electromagnetic fields, with their electric (E) and magnetic (H) components, plays a crucial role in both wireless power transfer and energy harvesting. Radiating power via technologies like microwave transmission offers potential for long-distance energy transfer but faces challenges in efficiency and environmental impact.

Combining a deeper understanding of electromagnetic principles with continued advancements in energy harvesting and green energy solutions holds immense promise for powering a more sustainable and interconnected world. By optimizing energy transfer methods and diversifying our energy mix, we can create a future where clean and efficient energy powers our homes, industries, and technologies.

REFERENCES

1. Balanis, C. A. (2012). *Advanced engineering electromagnetics*. John Wiley & Sons.
2. Pozar, D. M. (2009). *Microwave engineering*. John Wiley & Sons.
3. Rappaport, T. S., Seidel, S. Y., & Heath Jr., R. W. (1999). Millimeter wave wireless communications. *IEEE Transactions on Microwave Theory and Techniques, 47*(12), 2375–2389.
4. Karma, N. H., & Le Blanc, M. A. (2004). Mechanism of radiofrequency heating in neoplasia. *IEEE Transactions on Biomedical Engineering, 51*(1), 147–155
5. Mudgett, I. (1988). Microwave applications in food processing. *Food Technology, 42*(1), 76–86.
6. Ulaby, F. T., & Long, D. G. (2014). *Microwave remote sensing fundamentals and radiometric measurements*. Artech House
7. Born, M., & Wolf, E. (1999). *Principles of optics: electromagnetic theory of propagation, interference, and diffraction of light*. Cambridge University Press
8. Green, M. A. (2012). Solar cells: Converting sunlight into electricity. *Physica E: Low-dimensional Systems and Nanostructures, 48*(8), 1564–1572.
9. Priya, S., & Inman, D. J. (2009). *Energy harvesting technologies*. Springer Series in Materials Science and Engineering.
10. Roy, S. M., & Agrawal, D. P. (2014). An overview of ambient RF energy harvesting. *IEEE Circuits and Systems Magazine, 4*(2), 13–24.
11. Sudeval, S., Muñoz, P., Moreu, A., Xing, D., & Xhafa, F. (2013). Energy harvesting technologies for self-powered wireless sensor nodes. *Sensors (Basel), 13*(12), 16752–16786.
12. Mukhopadhyay, S. C. (2014). Wearable sensors for human health monitoring. *Medical Measurements and Metrics, 4*(1), 2.
13. Nigmatullin, D. E., et al. (2020). Energy harvesting for internet of things: review of technologies and applications. *Sensors (Basel)*, 20(18), 5400.

Chapter 3

Radiation in real life

RADIATION IN REAL LIFE

Radiation, a fundamental aspect of our universe, encompasses the transfer of energy through space and matter in the form of waves or particles. Radiation is a pervasive force in everyday life, influencing everything from natural phenomena to technological advancements. Solar radiation is essential for life, driving photosynthesis and climate patterns. Medical technologies like X-rays and radiation therapy use controlled doses for diagnosis and treatment, while everyday exposure to UV rays and electromagnetic fields from devices raises health considerations. Industrial applications and environmental monitoring also rely on radiation for safety and efficiency. Balancing its beneficial uses with safety measures is crucial to harnessing radiation responsibly and ensuring public health.

Early discoveries and the electromagnetic spectrum

The study of radiation has a rich history, with early observations dating back to ancient civilizations. The Greeks noted the attractive properties of amber (elektron) when rubbed with fur, a phenomenon later understood as electricity. In the 17th century, Isaac Newton observed the dispersion of sunlight through a prism, demonstrating the existence of a spectrum of colors. However, it was James Clerk Maxwell who, in the 19th century, unified the understanding of electricity and magnetism with his groundbreaking theory of electromagnetism. This theory predicted the existence of electromagnetic waves, which could travel through space at the speed of light. This prediction was later confirmed by Heinrich Hertz in 1887, marking a turning point in our understanding of radiation.

The electromagnetic spectrum is a continuous range of electromagnetic waves categorized by their wavelength and frequency. It encompasses a vast spectrum, ranging from low-frequency radio waves to high-frequency gamma rays. Visible light, with wavelengths detectable by the human eye, occupies a tiny portion of this spectrum.

DOI: 10.1201/9781003590712-3

Natural sources of radiation

Nature is a prolific source of radiation [1] across the electromagnetic spectrum. Here are some prominent examples:

Cosmic background radiation: This faint afterglow from the Big Bang permeates the universe and falls within the microwave range of the spectrum.

Solar radiation: The sun emits a broad spectrum of radiation, including visible light, ultraviolet (UV) radiation, and infrared (IR) radiation.

Radioactive decay: Unstable atomic nuclei spontaneously emit radiation in the form of alpha particles, beta particles, and gamma rays during a process known as radioactive decay.

These natural sources of radiation have been a constant presence on Earth and play a crucial role in various environmental processes. For instance, solar radiation fuels life on Earth through photosynthesis, while cosmic rays contribute to the formation of new elements.

Man-made sources of radiation

Human ingenuity [2] has also led to the development of various sources of radiation:

Radio waves and microwaves: These are utilized in various applications like radio and television broadcasting, mobile communication networks, and microwave ovens.

X-rays: High-energy X-rays are used for medical imaging and diagnostic purposes.

Gamma rays: Gamma rays are emitted by radioactive materials and are used in sterilization, cancer treatment, and industrial processes.

Nuclear reactors: Nuclear fission reactions in power plants generate electricity while also producing ionizing radiation, primarily neutrons and gamma rays.

These man-made sources of radiation offer significant benefits but also warrant careful management to minimize potential risks.

Applications of radiation

Radiation plays a vital role in diverse fields, offering numerous applications:

Medical applications: X-rays and gamma rays are crucial for medical imaging and cancer treatment. Radioisotopes are used in diagnostic tests and treatments for various medical conditions.

Communication technologies: Radio waves and microwaves form the backbone of wireless communication, enabling radio, television, mobile phones, and internet connectivity.

Non-destructive testing: X-rays and other forms of radiation are used for non-destructive testing to inspect materials for defects or anomalies, ensuring safety and quality control in various industries.

Scientific research: Radiation across the spectrum is used in various scientific research fields, from astronomy studying distant celestial objects to material science analyzing the structure of materials.

The applications of radiation have revolutionized various aspects of our lives, enabling advancements in healthcare, communication, and scientific understanding.

Risks associated with radiation exposure

While radiation offers numerous benefits, excessive exposure can pose health risks. The severity of these risks depends on the type, intensity, and duration of exposure. Ionizing radiation, such as X-rays, gamma rays, and alpha particles, has the potential to damage living tissues and increase the risk of cancer. Managing radiation exposure is crucial, particularly in medical applications, industrial settings, and nuclear power generation. This involves implementing safety measures like shielding, minimizing exposure times, and monitoring radiation levels to ensure worker and public safety.

The future of radiation research and technology

The field of radiation research continues to evolve, with ongoing efforts to:

- Develop more efficient and targeted radiation therapies for cancer treatment.
- Explore the potential of nuclear fusion as a clean energy source.
- Improve radiation detection technologies for environmental monitoring and security applications.
- Advance our understanding of the effects of low-dose radiation exposure.

These advancements hold promise for safer and more effective applications of radiation in various fields. Additionally, research is ongoing to develop new materials with tailored radiation-shielding properties for enhanced protection in medical and industrial settings. Furthermore, space exploration necessitates a deeper understanding of cosmic radiation and its potential health risks on astronauts during long-duration missions. Developing effective shielding technologies for spacecraft and space habitats will be crucial for ensuring astronaut safety in the vast expanse of space.

FREQUENCY BANDS

The electromagnetic spectrum, a vast and invisible ocean of energy, encompasses a wide range of frequencies. Understanding these frequencies and their distinct properties is fundamental to various technologies that permeate our daily lives.

Demystifying the electromagnetic spectrum

Imagine a light wave. Its color, perceived by the human eye, corresponds to a specific frequency and wavelength. The electromagnetic spectrum is a continuous range of these waves, encompassing frequencies from extremely low (ELF) to incredibly high (gamma rays). Each frequency band exhibits unique properties and interacts with matter differently.

Here's a helpful analogy: Think of the spectrum as a piano keyboard. Each key represents a specific frequency, with the lower keys corresponding to lower frequencies and longer wavelengths, and the higher keys representing higher frequencies and shorter wavelengths.

Frequency bands: a spectrum divided

The electromagnetic spectrum is further subdivided into distinct frequency bands based on their characteristics and applications. Here's an exploration of some prominent bands as seen in Figure 3.1:

Extremely low frequency (ELF): (0–30 Hz)—These low-frequency waves can penetrate deep into the Earth and water, making them suitable for applications like submarine communication and geophysical exploration.

Low frequency (LF): (30 Hz–300 kHz)—LF waves can travel long distances following the curvature of the Earth. They are used in radio navigation systems, AM radio broadcasting, and geophysical research.

Medium frequency (MF): (300 kHz–3 MHz)—MF waves have a shorter range compared to LF but are still effective for regional radio broadcasting, particularly AM radio.

High frequency (HF): (3 MHz–30 MHz)—HF waves can reflect off the ionosphere, enabling long-distance communication over shortwave radio, amateur radio, and citizen band (CB) radio.

Very high frequency (VHF): (30 MHz–300 MHz)—VHF waves offer shorter range communication but are less susceptible to ionospheric variations. They are used in FM radio broadcasting, marine communication, aviation communication, and two-way radios.

Ultra high frequency (UHF): (300 MHz–3 GHz)—UHF waves provide line-of-sight communication and are utilized in television

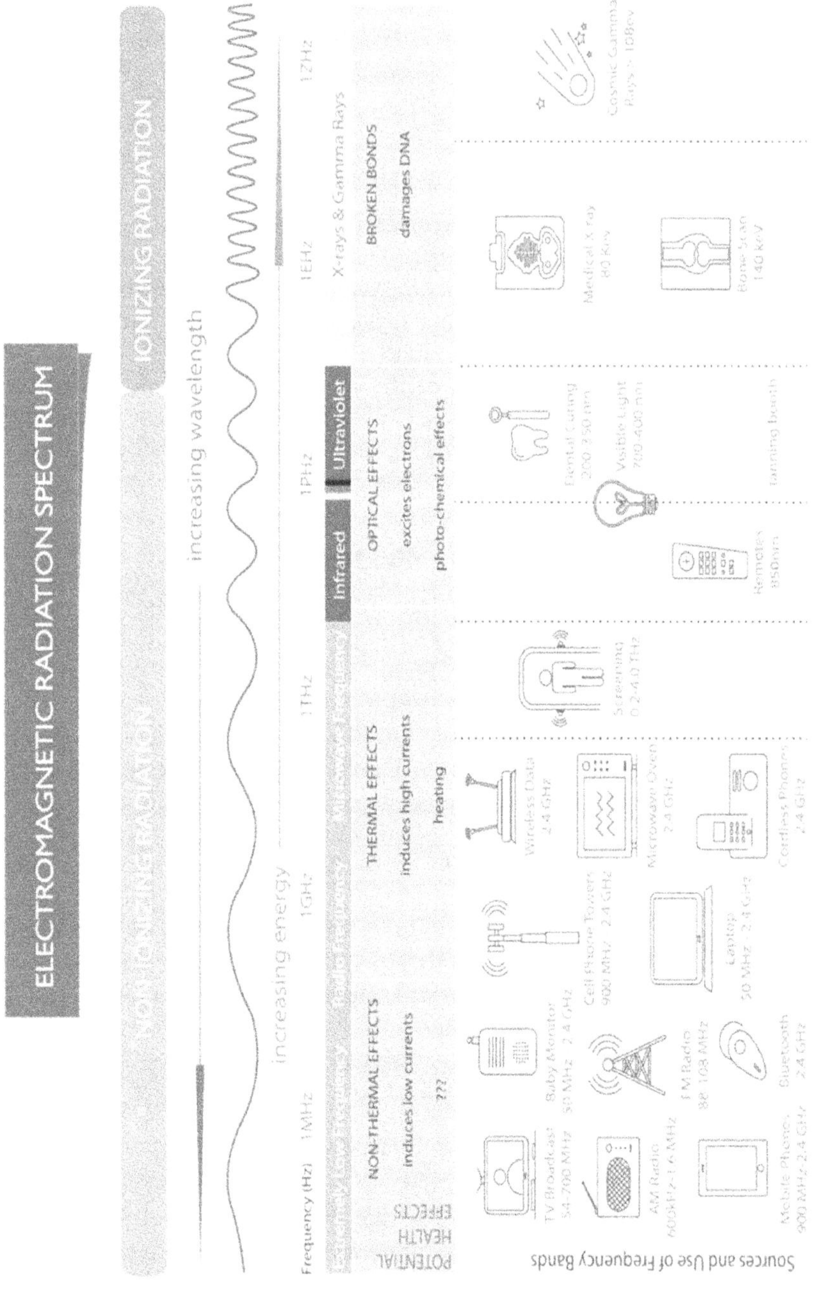

Figure 3.1 Frequency bands

broadcasting, mobile phone networks (4G and 5G), satellite communication, and radar systems.

Super high frequency (SHF): (3 GHz–30 GHz)—SHF waves have even shorter wavelengths and higher frequencies. They are used in satellite communication, radar systems, microwave ovens, and some wireless networking technologies.

Extremely high frequency (EHF): (30 GHz–300 GHz)—EHF waves are used in radar systems, millimeter-wave communication technologies, and security scanners.

Infrared (IR): (300 GHz–430 THz)—Infrared radiation is invisible to the human eye but detectable as heat. It has applications in night vision technology, thermal imaging, remote controls, and medical diagnostics.

Visible light: (430 THz–750 THz)—This narrow band within the spectrum corresponds to the colors we perceive, from red to violet.

Ultraviolet (UV): (750 THz–30 PHz)—Ultraviolet radiation is invisible to the human eye and can cause sunburn. It has applications in sterilization, water purification, and fluorescent lighting.

X-rays: (30 PHz–30 EHz)—High-energy X-rays are used for medical imaging, security scanners, and industrial radiography.

Gamma rays: (Above 30 EHz)—Gamma rays are the most energetic form of electromagnetic radiation and are used in cancer treatment, sterilization, and astronomical research.

Applications of frequency bands in everyday life

Communication: Wi-Fi and Bluetooth utilize specific UHF and SHF bands for short-range wireless communication between devices like laptops, smartphones, and speakers.

Navigation: GPS (Global Positioning System) relies on high-precision satellite signals transmitted in the L-band (1–2 GHz) for accurate location tracking.

Safety and security: Radar systems operating in various bands (UHF, SHF) provide vital detection and tracking capabilities for air traffic control, weather monitoring, and maritime navigation. Security scanners at airports utilize millimeter-wave technology (EHF) for object detection.

Space exploration: Communication with spacecraft and deep-space probes relies on various frequency bands depending on distance and data transmission requirements.

Medical applications: X-ray imaging utilizes high-energy X-rays to produce detailed pictures of bones and internal structures. Medical imaging techniques like MRI (Magnetic Resonance Imaging) also involve specific radio frequencies to manipulate and detect the magnetic properties of atoms within the body.

Scientific research: Radio astronomy observes distant celestial objects at various frequencies across the spectrum, revealing unique information about their composition and behavior.

Industrial applications: Microwave heating in industrial processes utilizes specific frequencies to heat materials efficiently, used in food processing and drying applications. Additionally, non-destructive testing techniques like radiography employ X-rays or gamma rays to inspect materials for defects.

These are just a few examples, and the applications of frequency bands continue to evolve as technology advances.

Challenges and considerations with frequency spectrum usage

Managing the electromagnetic spectrum effectively is crucial to avoid interference between different users. Here are some key challenges:

- *Spectrum scarcity*: The demand for wireless communication technologies is constantly increasing, leading to a scarcity of available spectrum resources.
- *Spectrum allocation and regulation*: Regulatory bodies need to efficiently allocate spectrum licenses to different users while ensuring fair competition and preventing interference.
- *Technological advancements*: New technologies often require new frequency bands, necessitating dynamic allocation strategies and spectrum refarming techniques.
- *Environmental considerations*: Certain frequency bands can be affected by natural phenomena like solar flares, and potential environmental impacts of high-power transmission systems need to be carefully assessed.

The future of frequency bands: innovation and efficiency

The future of frequency bands hinges on innovation and efficient spectrum utilization:

Cognitive radio technology: These intelligent radios can dynamically adjust their operating frequency based on availability, reducing interference and optimizing spectrum usage.

Spectrum sharing techniques: Sharing spectrum between different users in a non-intrusive manner requires sophisticated techniques like dynamic spectrum access and spectrum overlays.

Millimeter-wave and terahertz technologies: Exploring higher frequency bands like millimeter-wave and terahertz holds promise for ultra-fast wireless communication applications.

Satellite communication advancements: New constellations of smaller satellites promise greater capacity and flexibility in data transmission across the globe.

By actively addressing these challenges and promoting technological advancements, we can ensure efficient and sustainable use of the electromagnetic spectrum for a connected and innovative future.

AFFECTING FLORA AND FAUNA

Radiation, the invisible transfer of energy through space and matter, is a ubiquitous force in our environment. While often associated with human activities like nuclear power generation or medical treatments, radiation also exists naturally in the form of cosmic rays and terrestrial radioactivity. Living organisms have always coexisted with natural background radiation. Plants and animals have evolved various mechanisms to mitigate the effects of natural background radiation. However, human activities can significantly alter these natural levels, potentially impacting ecosystems [3].

Exploring the effects of radiation on plants

Plants exhibit a wide range of sensitivity to radiation. Some key considerations include:

Growth and development: High doses of radiation can stunt growth, reduce seed germination, and inhibit flowering.

Genetic mutations: Radiation can cause mutations in plant DNA, potentially leading to heritable changes and phenotypic variations.

Physiological processes: Radiation can disrupt photosynthesis, nutrient uptake, and other vital physiological processes in plants.

However, plants also demonstrate a degree of resilience:

DNA repair mechanisms: Plants possess DNA repair mechanisms that can help mitigate damage caused by low-dose radiation exposure.

Acclimation: Some plant species can adapt to chronic low-dose radiation exposure by modifying their cellular processes.

Differential sensitivity: Different plant species and tissues exhibit varying degrees of sensitivity to radiation.

The impact of radiation on plant populations depends on the type, dose, and duration of exposure, along with the specific plant species and its ecological context.

Unveiling the impact on animals

Animals, similar to plants, exhibit diverse responses to radiation [4]. Here are some areas to consider:

Acute radiation syndrome (ARS): High doses of radiation can cause ARS, a complex illness affecting multiple organ systems, leading to death in severe cases.

Increased cancer risk: Radiation exposure can increase the risk of cancer development in animals, depending on the dose and type of radiation.

Reproductive effects: Radiation can affect fertility, embryo development, and offspring survival.

Genetic effects: Mutations caused by radiation can be passed on to future generations, potentially impacting population health.

However, resilience is also observed in the animal kingdom:

DNA repair mechanisms: Animals possess DNA repair mechanisms, similar to plants, to mitigate radiation damage.

Behavioral adaptations: Some animal species exhibit behavioral adaptations like avoiding areas with high radiation levels.

Varying sensitivity: Different animal species and life stages exhibit diverse levels of sensitivity to radiation.

The ecological impact of radiation on animal populations depends on factors like species diversity, habitat complexity, and overall radiation exposure levels.

Case studies: radiation's impact on ecosystems

Several real-world scenarios illustrate the potential consequences of radiation exposure on ecosystems:

Chernobyl nuclear disaster: The 1986 Chernobyl disaster caused widespread radioactive contamination, impacting plant and animal populations within the exclusion zone. While immediate effects were severe, long-term studies reveal ongoing adaptations and population recovery in some species.

Fukushima Daiichi nuclear disaster: The 2011 Fukushima Daiichi nuclear accident resulted in radioactive releases into the environment. Studies are ongoing to assess the long-term effects on marine ecosystems and terrestrial wildlife populations.

Nuclear weapons testing: Past nuclear weapons testing has left radioactive legacies in some environments. These areas offer unique opportunities to study the long-term effects of radiation exposure on flora and fauna.

These case studies highlight the importance of long-term monitoring and ecological research to understand the complex interplay between radiation exposure, population dynamics, and ecosystem health.

Mitigating risks and fostering environmental protection

Minimizing the potential risks of radiation on flora and fauna requires a multi-pronged approach:

Strict environmental regulations: Stringent regulations on radioactive waste disposal and nuclear power plant operations are crucial to prevent environmental contamination.

Radioactive material management: Safe handling, storage, and disposal of radioactive materials are essential to limit unintended releases into the environment.

Habitat protection: Conservation efforts for sensitive ecosystems near potential radiation sources can help mitigate the impact on vulnerable species.

Environmental monitoring: Continuous monitoring of radiation levels in the environment is essential to detect any potential contamination and track its impact on flora and fauna.

Research and development: Continued research on radiation's effects at the cellular and organismal level can inform better environmental protection strategies.

By implementing these measures, we can strive for a balance between harnessing the potential of nuclear technologies and minimizing the environmental impact on flora and fauna. By recognizing the potential risks associated with radiation exposure, implementing robust safety measures, and fostering responsible practices, we can strive for a future where humans and the natural world coexist safely with this invisible force. Continued research holds the key to unlocking a deeper understanding of radiation's impact on living organisms and developing effective strategies for mitigation and adaptation.

AFFECTING NEWBORN AND AGE OLD

While the benefits of certain radiation-based technologies are undeniable, understanding their potential harm to the nervous system across different life stages is crucial. The developing nervous system in newborns [5] is particularly vulnerable to radiation exposure. Some key concerns include:

Disrupted brain development: High doses of ionizing radiation, particularly during the first trimester of pregnancy or shortly after birth, can hinder brain development, potentially leading to cognitive impairments and learning difficulties.

Increased cancer risk: Radiation exposure in early life can elevate the risk of childhood cancers, especially those affecting the brain and central nervous system.

Neurological disorders: While the exact mechanisms are still under investigation, some studies suggest a potential link between prenatal radiation exposure and the development of neurological disorders like autism spectrum disorder (ASD).

However, it's important to note that these risks are often associated with high-dose exposures, such as those received during medical procedures or accidental radiation incidents. Minimizing unnecessary radiation exposure during pregnancy and infancy remains a crucial public health focus.

The elderly nervous system [6] also exhibits increased vulnerability to radiation exposure due to several factors:

Reduced repair mechanisms: With age, the body's natural ability to repair DNA damage caused by radiation diminishes, potentially leading to long-term health consequences.

Pre-existing conditions: Elderly individuals often have pre-existing neurological conditions like dementia or Alzheimer's disease. Radiation exposure may exacerbate these conditions, accelerating cognitive decline.

Increased sensitivity: Certain medications commonly used by the elderly can potentially increase their sensitivity to radiation's effects.

The nervous system in the elderly is particularly susceptible to the following radiation-related issues:

Cognitive decline: Exposure to radiation, even at moderate levels, may accelerate cognitive decline and increase the risk of dementia in the elderly population.

Balance and coordination problems: Radiation damage to the nervous system can impair balance and coordination, potentially leading to falls and injuries.

Increased stroke risk: Some studies suggest a potential link between radiation exposure and an elevated risk of stroke in the elderly.

Pregnant women should discuss potential radiation risks with their healthcare providers and avoid unnecessary medical imaging procedures during the first trimester if possible. Healthcare professionals must ensure that the potential benefits of any radiation-based medical procedures for both newborns and the elderly outweigh the risks, exploring alternative diagnostic methods when feasible. Proper shielding techniques during medical procedures can significantly reduce radiation exposure to sensitive areas, particularly the head and brain in newborns. Raising awareness about the potential risks of radiation exposure and promoting safe practices among caregivers of newborns and the elderly is crucial. Additionally, continued research on the long-term effects of radiation exposure on the nervous system across different age groups is essential for developing effective mitigation strategies.

PERMISSIBLE LIMITS FOR HUSTLE-FREE LIVING CONDITIONS

The idea of hustle-free living includes several key elements. Minimizing exposure to environmental stressors like air pollution, noise, and excessive light disruptions reduces stress. Maintaining agreeable temperature and humidity levels supports optimal physiological performance and sleep quality. Thoughtful lighting and design create an environment that promotes rest and lowers cognitive stress. Improving air quality and lighting enhances concentration and encourages original thought. While achieving the ideal balance is challenging, setting acceptable boundaries for these factors provides a framework for fostering a positive and productive atmosphere.

Noise pollution and its permissible limits

Noise pollution [7] is a significant environmental stressor impacting health and well-being. Here's a breakdown of permissible noise limits established by various organizations:

World Health Organization (WHO): Recommends daytime noise levels below 53 decibels (dB) Lden (Day-Evening-Night) for good sleep quality.

European Union (EU): Sets a daytime limit of 55 dB Lden and a nighttime limit of 50 dB Lden for residential areas.

United States Environmental Protection Agency (USEPA): Recommends a continuous daytime limit of 70 dB outside residences and 55 dB inside residences for conversations and activities without strain.

Technical considerations

Decibel (dB): The unit for measuring sound pressure level.

Lden: Day-Evening-Night noise level, a metric accounting for variations in noise levels throughout the day.

Noise reduction coefficient (NRC): A rating system for building materials, indicating their sound absorption ability.

Air quality standards for a healthy environment

Air quality plays a crucial role in respiratory health and overall well-being. Here are key pollutants and their permissible limits:

Particulate matter (PM): The WHO recommends annual average PM2.5 (fine particulate matter) levels below 10 $\mu g/m^3$ and PM10 (coarse particulate matter) levels below 45 $\mu g/m^3$.

Ozone (O₃): The USEPA sets an 8-hour average ozone limit of 0.070 parts per million (ppm).

Nitrogen dioxide (NO₂): The WHO recommends an annual average NO_2 limit of 10 $\mu g/m^3$.

Technical considerations

Micrograms per cubic meter (µg/m³): Unit for measuring the mass concentration of a pollutant in air.

Parts per million (ppm): Unit for measuring the concentration of a pollutant relative to a million molecules of air.

Air quality monitoring systems: Continuously monitor air quality by measuring pollutant concentrations and providing real-time data.

Light and color: optimizing for comfort and productivity

Lighting plays a significant role in regulating our circadian rhythm, impacting sleep, mood, and alertness. Here are some considerations for permissible light levels:

Daytime: Maintain illuminance levels between 300 and 500 lux for most tasks, with higher levels (up to 1000 lux) for visually demanding activities.

Night time: Aim for dim lighting below 50 lux to promote melatonin production and sleep quality.

Color temperature: Warm white light (around 2700 Kelvin) is generally considered calming, while cooler white light (around 5000 Kelvin) can promote alertness.

Technical considerations

Lux: Unit for measuring the illuminance level, which is the amount of light falling on a surface.

Kelvin (K): Unit for measuring color temperature, with lower Kelvin values representing warmer light and higher values representing cooler light.

Light emitting diodes (LEDs): LED bulbs offer energy efficiency and tunable color temperature options, facilitating the creation of customized lighting environments.

Thermal comfort: maintaining a balanced environment

Thermal comfort refers to the state of mind where a person feels neither too hot nor too cold. Here's a breakdown of permissible temperature and humidity ranges for a comfortable and healthy environment:

Temperature: The American Society of Heating, Refrigerating and Air-Conditioning Engineers (ASHRAE) recommends a comfort range of 22–25.5°C (71.6–77.9°F) for occupants wearing typical office clothing.

Humidity: A relative humidity range of 30–60% is generally considered comfortable. Lower humidity can cause dry skin and respiratory issues, while higher humidity can promote mold growth.

Technical considerations

Dry-bulb temperature: The air temperature measured by a standard thermometer.

Relative humidity: The percentage of moisture in the air compared to the maximum amount it can hold at that temperature.

Thermal comfort standards: Organizations like ASHRAE establish comfort standards based on factors like temperature, humidity, air velocity, and clothing insulation.

HVAC systems: Heating, ventilation, and air conditioning systems help maintain desired temperature and humidity levels within a building.

The concept of a hustle-free living environment goes beyond the absence of stress; it encompasses a sense of security and safety. By implementing responsible practices regarding radiation use in medical procedures, industrial applications, and waste management, we can strive to minimize potential health risks and anxieties associated with radiation exposure. Living in harmony with the natural world, including its inherent radiation background, involves a balance between harnessing the benefits of technology and safeguarding our health. As we continue to explore the intricate connections between radiation and living organisms, this awareness empowers us to make informed choices and promote sustainable practices that contribute to a future where well-being and a "hustle-free" existence are attainable for all.

DISEASES CAUSED AND CURES AVAILABLE

Ionizing radiation can induce various adverse health effects, with the severity depending on the factors mentioned above. Following are some key diseases associated with radiation exposure:

Acute radiation sickness (ARS): This syndrome occurs following high-dose exposure (typically exceeding 1 Gy) within a short period, causing damage to rapidly dividing cells in the gastrointestinal tract, bone marrow, and skin. Symptoms can range from nausea and vomiting to hair loss, internal bleeding, and death in severe cases.

Medical management: Supportive care often forms the cornerstone of treatment for ARS, including fluid resuscitation, blood product transfusions, and infection management. Bone marrow transplants may be considered in cases with severe bone marrow damage.

Cancer: Radiation exposure is a known carcinogen, increasing the risk of developing various cancers, including leukemia, thyroid cancer, lung cancer, and skin cancer. The risk increases with higher doses and varies depending on the type of radiation and the exposed tissue.

Cancer treatment options: Treatment approaches for radiation-induced cancers are similar to those for cancers arising from other causes and may include surgery, chemotherapy, radiation therapy (ironically using controlled radiation doses to target cancerous cells), and targeted therapies.

Genetic mutations: Radiation can damage DNA, leading to mutations that can be passed on to future generations. The long-term consequences of these mutations are still under investigation.

Genetic counseling: Individuals with a history of significant radiation exposure may benefit from genetic counseling to understand the potential risks for themselves and their offspring.

Non-malignant diseases: Chronic radiation exposure, even at relatively low doses, can contribute to the development of non-malignant diseases like cataracts, cardiovascular diseases, and certain lung conditions.

Preventative measures: Limiting unnecessary exposure and implementing radiation safety protocols in occupational settings are crucial for minimizing the risk of these diseases.

The challenge of cures: managing radiation-induced diseases

Unfortunately, there aren't specific "cures" for radiation-induced diseases. Management strategies primarily focus on:

Supportive care: This involves managing symptoms and providing treatments to improve the patient's quality of life. This can include pain management, nutritional support, and psychological counseling.

Treating specific diseases: For cancers, conventional treatment options like surgery, chemotherapy, and radiation therapy come into play.

Minimizing further exposure: Preventing additional radiation exposure is crucial for promoting healing and reducing the risk of further complications.

Mitigating risks and fostering a safer future

Several measures can be implemented to mitigate the risks associated with radiation exposure:

Dose optimization: In medical procedures and other applications, optimizing radiation doses to achieve desired results while minimizing exposure is crucial. Medical imaging techniques like ultrasound and magnetic resonance imaging (MRI) can sometimes be employed as alternatives to X-rays or CT scans, reducing radiation burden.

Shielding techniques: Utilizing physical barriers like lead aprons and walls in medical settings can significantly reduce exposure for patients and healthcare professionals. Similarly, in occupational settings where radiation exposure is a concern, appropriate shielding materials and safety protocols are essential.

Safety regulations: Regulatory bodies like the International Commission on Radiological Protection (ICRP) and national agencies establish guidelines for safe radiation exposure limits in workplaces, medical settings, and environmental contexts. These regulations play a crucial role in protecting public health.

Technical considerations

As Low As Reasonably Achievable (ALARA): This principle is a cornerstone of radiation safety, emphasizing minimizing radiation exposure to the lowest level achievable while still accomplishing the desired task.

Equivalent dose (Sv): This unit accounts for the different biological effects of various radiation types, allowing for a standardized comparison of radiation exposure risks.

Dosimetry: This field focuses on measuring radiation dose [8] absorbed by living organisms. Various detectors and techniques are employed, like ionization chambers, thermoluminescent dosimeters (TLDs), and semiconductor detectors. Understanding dosimetry is crucial for implementing effective dose optimization strategies.

EMERGING FRONTIERS: RESEARCH AND ADVANCEMENTS IN RADIATION ONCOLOGY

The field of radiation oncology is constantly evolving, exploring novel means to utilize radiation for cancer treatment while minimizing side effects. Here are some promising areas of research:

Targeted radionuclide therapy: This approach employs radioactive isotopes that specifically target cancer cells, delivering a concentrated dose of radiation while minimizing damage to healthy tissues.

Proton therapy and heavy ion therapy: These advanced techniques utilize charged particles (protons or heavier ions) that deliver their dose more precisely to tumor sites, reducing damage to surrounding tissues compared to conventional X-ray therapy.

Radiomics: This emerging field combines medical imaging data with advanced computational analysis to predict tumor response to radiation therapy and personalize treatment plans.

Radiation plays a complex role in our world, offering both benefits and risks. By understanding the types of radiation, their impact on the human body, and the potential diseases they can cause, we can implement strategies to minimize exposure and manage its health consequences. Continued research in radiation safety practices, advanced treatment techniques, and a focus on dose optimization are crucial for fostering a safer future where the benefits of radiation can be harnessed while minimizing its potential harm.

REFERENCES

1. United Nations Scientific Committee on the Effects of Atomic Radiation. (1993). Exposures from natural sources of radiation. In *Report of the United Nations Scientific Committee on the Effects of Atomic Radiation* (pp. 33–89). https://doi.org/10.18356/c3a60b10-en

2. English, R. A., Benson, R. E., Bailey, J. V., & Barnes, C. M. (1973). *Apollo experience report: protection against radiation.* https://ntrs.nasa.gov/api/citations/19730010172/downloads/19730010172.pdf

3. Fukumoto, M. (2019). *Low-dose radiation effects on animals and ecosystems.* Springer Nature.

4. Eisler, R. (1994). *Radiation hazards to fish, wildlife, and invertebrates.* U.S. Department of the Interior, National Biological Service.

5. Moon, J. H. (2020). Health effects of electromagnetic fields on children. *Clinical and Experimental Pediatrics, 63*(11), 422–428. https://doi.org/10.3345/cep.2019.01494

6. Hunter, C. P., Johnson, K. A., & Muss, H. B. (2000). *Cancer in the elderly.* CRC Press.

7. Peñafiel, P., Cazares, K., & Marizande, D. (2018). Evaluation of the noise pollution of the Cevallos avenue, Ambato city, Ecuador, for the determination of critical points of affectation in the place, and to provide information that allows the generation of preventive and corrective measures to the problem. *Proceedings of MOL2NET 2018, International Conference on Multidisciplinary Sciences*, 4th Edition. https://doi.org/10.3390/mol2net-04-05912

8. Nath, R., Anderson, L. L., Luxton, G., Weaver, K. A., Williamson, J. F., & Meigooni, A. S. (1995). Dosimetry of interstitial brachytherapy sources: recommendations of the AAPM Radiation Therapy Committee Task Group No. 43. *Medical Physics, 22*(2), 209–234. https://doi.org/10.1118/1.597458

Chapter 4

Environmental impact

EMISSIONS AND WEATHER

The price of progress: emissions and the disruption of weather

Humanity's relentless pursuit of technological advancement has undeniably propelled us forward. Yet, this progress often comes at a hidden cost—the degradation of our environment. The emissions we generate through various technological processes significantly impact weather patterns, seasons, and ultimately, the delicate balance of our planet.

A spectrum of emissions: from industry to agriculture

Technological advancements fuel our industries, transportation systems, and even agriculture. Each sector contributes a unique blend of emissions to the atmosphere:

- *Industrial emissions*: Factories and power plants heavily rely on fossil fuels like coal, oil, and natural gas for energy. The burning of these fuels releases a cocktail of pollutants, including carbon dioxide (CO_2), nitrogen oxides (NO_x), and sulfur oxides (SO_x), all potent greenhouse gases that trap heat in the atmosphere, contributing to global warming. A 2023 report by the International Energy Agency (IEA) revealed that the industrial sector remains the largest single emitter of CO_2 globally, accounting for over a quarter of total emissions [1].
- *Transportation emissions*: Vehicles powered by gasoline and diesel engines are significant contributors of air pollution. They emit CO_2, NO_x, and particulate matter (PM), a microscopic dust-like substance linked to respiratory illnesses. A study by the Environmental Protection Agency (EPA) in the United States found that transportation was the leading source of greenhouse gas emissions in the country in 2020, contributing to over 29% of total emissions [2].

DOI: 10.1201/9781003590712-4

- *Agricultural emissions*: Modern agricultural practices also play a role in emissions. Intensive farming methods, relying heavily on fertilizers, often lead to the release of nitrous oxide (N_2O), a powerful greenhouse gas with nearly 300 times the heat-trapping potential of CO_2. Additionally, livestock raised for meat production emit methane, another potent greenhouse gas, through their digestive processes. A 2019 report by the Food and Agriculture Organization (FAO) of the United Nations estimates that agriculture, forestry, and other land-use activities contribute to nearly 24% of global greenhouse gas emissions [3].

The ripple effect: from warming trends to weather extremes

These emissions have a profound impact on our weather patterns. The primary effect is global warming, a gradual increase in Earth's average temperature due to the greenhouse effect. This warming disrupts natural climate cycles, leading to a cascade of consequences:

- *Extreme weather events*: A warmer planet creates a more energetic atmosphere, leading to an increase in the intensity and frequency of extreme weather events. This includes heat waves, droughts, floods, hurricanes, typhoons, and wildfires. For instance, a 2022 study published in Nature Geoscience linked the devastating heatwaves scorching Europe in 2019 to human-induced climate change [4].
- *Shifting seasons*: As global temperatures rise, the traditional patterns of seasons are disrupted. Spring arrivals may be earlier, winters milder, and autumns shorter. This can have cascading effects on ecosystems, impacting plant and animal life cycles. A 2021 study in the Proceedings of the National Academy of Sciences (PNAS) documented a significant shift in spring arrival dates for plant life across North America, with some regions experiencing earlier blooming by as much as two weeks [5].
- *Rising sea levels*: Melting glaciers and polar ice caps due to global warming contribute to rising sea levels, threatening coastal communities and ecosystems. A 2023 report by the Intergovernmental Panel on Climate Change (IPCC) projects that sea levels could rise by up to one meter by 2100 under a high-emissions scenario, displacing millions of people worldwide [6].

Real-world stories: a glimpse into the future

The consequences of disrupted weather patterns are no longer theoretical; they are unfolding across the globe:

- *Pakistan's devastating floods*: In 2023, unprecedented monsoon rains caused catastrophic flooding in Pakistan. This event, linked to climate change by scientists, submerged a third of the country, displacing millions and causing billions of dollars in damage [7].
- *California's wildfire woes*: California has witnessed an alarming increase in the frequency and intensity of wildfires in recent years. These fires, fueled by drier conditions due to climate change, have devastated vast swathes of land and forced mass evacuations [8].
- *The disappearing Arctic sea ice*: Arctic sea ice cover has been declining at an alarming rate. This not only threatens polar ecosystems but also disrupts weather patterns in the Northern Hemisphere, potentially leading to more extreme winters in Europe and North America [9].

Moving forward: toward a sustainable future

The current situation demands immediate action. Transitioning to cleaner technologies and adopting sustainable practices are crucial steps toward mitigating the impact of emissions on our weather patterns. Here are some potential solutions:

Shifting to renewable energy: Replacing fossil fuels with renewable energy sources like solar, wind, geothermal, and hydropower can significantly reduce greenhouse gas emissions. Governments and industries must invest in renewable energy infrastructure and incentivize its adoption by consumers.

Promoting electric vehicles: Transportation electrification plays a vital role in curbing emissions. Investing in charging infrastructure and promoting electric vehicles can significantly reduce transportation-related emissions. Additionally, developing cleaner fuel alternatives like hydrogen fuel cells holds promise for the future.

Sustainable agriculture: Transforming agricultural practices toward regenerative farming methods can significantly reduce emissions and improve soil health. This includes practices like crop rotation, cover cropping, and composting, which can help sequester carbon in the soil and reduce reliance on synthetic fertilizers.

Global cooperation: Addressing climate change requires a global effort. International cooperation is essential for developing and implementing effective policies to reduce emissions and mitigate the effects of climate change.

Individual action matters

While large-scale changes are necessary, individual actions also contribute. Here's how individuals can make a difference:

Reduce energy consumption: Conserving energy at home by using energy-efficient appliances and adopting practices like turning off lights and electronics when not in use can significantly reduce emissions.

Travel smart: Opting for public transportation, cycling, or walking whenever possible can significantly reduce transportation emissions. Additionally, carpooling or using fuel-efficient vehicles can help.

Reduce, reuse, recycle: Implementing the principles of a circular economy by reducing consumption, reusing items whenever possible, and recycling waste can minimize our environmental footprint.

Sustainable food choices: Reducing meat consumption and supporting local, sustainable food producers can contribute to lower emissions in the agricultural sector.

The connection between emissions and weather disruptions is undeniable. The evidence is clear—our relentless pursuit of progress has come at the cost of a changing climate. We are in a race against time to mitigate the effects of climate change before it's too late. By adopting cleaner technologies, transitioning to sustainable practices, and taking individual action, we can create a future where progress and environmental responsibility go hand in hand.

NATURAL DISASTERS

The fury of nature: how human greed fuels natural disasters

Natural disasters—earthquakes, floods, hurricanes, wildfires—have always been a part of Earth's history. These events shape landscapes, test human resilience, and remind us of the immense power of nature. However, in recent decades, the frequency and intensity of these disasters have seemingly increased. While the Earth's natural cycles play a role, a significant factor contributing to this rise is human activity.

The delicate balance of nature

Natural disasters occur due to disruptions in Earth's geophysical and atmospheric processes. Earthquakes happen due to the movement of tectonic plates; volcanoes erupt when molten rock builds up beneath the Earth's surface. Floods are caused by excessive rainfall, overflowing rivers, or storm surges. Hurricanes and typhoons form over warm ocean waters, drawing energy to become powerful storms. These events are often cyclical, with a natural balance in place. However, human activities are tipping the scales.

Human greed: a catalyst for catastrophe

Our insatiable desire for progress has consequences. Here's how human actions are fueling natural disasters:

Climate change: The burning of fossil fuels releases greenhouse gases, trapping heat in the atmosphere. This disrupts weather patterns, leading to more extreme weather events like floods, droughts, and heatwaves. A 2021 report by the Intergovernmental Panel on Climate Change (IPCC) linked human activity to the observed increase in global temperatures since the mid-20th century [10].

Deforestation: Forests play a vital role in regulating water flow and preventing soil erosion. However, deforestation for agriculture, logging, and development increases the risk of landslides and floods. A 2023 study published in *Nature Sustainability* found a clear link between deforestation and the intensity of floods in Southeast Asia [11].

Unsustainable practices: Intensive agricultural practices deplete soil nutrients and reduce its ability to retain water, leading to increased risk of droughts and floods. Additionally, improper waste disposal can clog drainage systems, worsening flooding in urban areas.

Real-world examples: nature's fury

The consequences of human actions are evident in real-life disasters:

Hurricane Harvey (2017): This devastating hurricane slammed into Texas, causing catastrophic flooding. Scientists linked the extreme rainfall, in part, to warmer ocean temperatures fueled by climate change [12].

Australian bushfires (2019–2020): Australia witnessed its worst wildfire season on record, fueled by prolonged droughts and extreme heat. This disaster, linked to climate change, caused widespread ecological damage and property loss [13].

Nepal earthquake (2015): This powerful earthquake devastated Nepal, highlighting the vulnerability of mountain regions to seismic activity. However, deforestation and poor construction practices contributed to the high death toll [14].

Curbing the fury: a multigenerational effort

While natural disasters are inevitable, mitigating their impact requires long-term solutions:

Transition to renewable energy: Shifting from fossil fuels to wind, solar, and geothermal energy can significantly reduce greenhouse gas emissions and slow down climate change.

Sustainable land management: Reforestation efforts, promoting soil conservation practices, and sustainable agriculture techniques are crucial for preventing soil erosion, regulating water flow, and reducing flood risks.

Disaster-resilient infrastructure: Building infrastructure that can withstand extreme weather events and earthquakes can minimize damage and loss of life.

International cooperation: The fight against climate change and natural disasters requires a global effort. International treaties and agreements are essential for coordinated action and resource sharing.

The increasing frequency and intensity of natural disasters serve as a stark warning. If we continue on our current path of unchecked resource consumption and environmental degradation, nature will inevitably take its course. Just as a pendulum swings back, so too will nature restore balance, potentially at our expense. Ecosystems will collapse, displacing and endangering countless species, including ourselves. Fertile lands will become barren wastelands, and coastal cities will be swallowed by rising seas. By curbing our environmental impact, supporting sustainable practices, and adopting long-term solutions, we can mitigate the fury of natural disasters and build a more resilient future for generations to come. Let us learn from the lessons of nature's power and ensure our advancements work in harmony with the delicate balance of our planet. Nature being vast and powerful, if we are not wise enough to respect its delicate balance, it will prove its overwhelming wisdom and restore all.

EXTREME WEATHER AREAS

Living on the edge

The Earth boasts a vast array of landscapes, with some regions naturally existing in harsh environments. From the scorching deserts of the Sahara to the frigid expanses of the Arctic, these areas present unique challenges to human and animal life. However, a new threat is emerging—the disruption of these delicate ecosystems due to a combination of extreme weather events and the ever-increasing presence of communication technologies.

Communication revolution: a double-edged sword

The rise of communication technologies like satellite internet presents a double-edged sword for these harsh environments.

Benefits: Improved communication can be a lifeline, facilitating better weather forecasting, emergency response coordination, and research efforts. For example, real-time weather data transmitted through satellite internet can help remote communities prepare for extreme weather events. Additionally, researchers studying polar ecosystems can use technology to monitor animal migration patterns and environmental changes.

Drawbacks: The infrastructure needed for communication networks, such as cell towers and data centers, can have a disruptive impact on these fragile ecosystems. Construction disrupts habitats, and the constant hum of electronic equipment can disrupt wildlife behavior. Additionally, increased human presence associated with improved communication can lead to increased resource use and pollution.

Extreme weather: a growing threat

Climate change is causing a rise in extreme weather events, further disrupting the delicate balance of harsh environments.

Melting glaciers: Rising temperatures are causing an alarming rate of glacial melt in polar regions. This not only contributes to sea level rise but also disrupts the delicate food webs that depend on these frozen ecosystems. A 2023 study published in *Nature Climate Change* documented the rapid retreat of glaciers in Greenland, contributing to a significant rise in global sea levels [15].

Intensified droughts: Areas like the Sahara are experiencing more frequent and intense droughts, further straining limited water resources. These droughts impact not only plant and animal life but also contribute to social unrest and displacement as communities struggle to survive. A 2022 report by the United Nations Office for Disaster Risk Reduction (UNDRR) highlighted the increasing frequency and severity of droughts in Africa, impacting millions of people [16].

Real-life examples: living on the edge

The Inuit of the Arctic: The Inuit people of the Arctic have long thrived in a harsh environment, relying on traditional hunting and fishing practices for survival. However, melting sea ice and unpredictable weather patterns are disrupting their way of life. A 2024 article in *The Guardian* documented the challenges faced by Inuit communities in Canada, with melting sea ice making traditional hunting practices dangerous [17].

The Tuareg of the Sahara: The nomadic Tuareg people of the Sahara have adapted to the desert's harsh conditions for centuries. However, recent droughts and sandstorms are forcing some communities to abandon their nomadic lifestyle and settle in overcrowded refugee camps. A 2022 report by Al Jazeera explored the plight of the Tuareg people, struggling with climate change and dwindling resources [18].

Finding a balance: toward sustainable solutions

The future of these extreme environments depends on our ability to manage technology and mitigate the effects of climate change:

Sustainable infrastructure: Developing communication technologies with minimal environmental impact is crucial. This includes exploring renewable energy sources for powering communication networks and using innovative designs to minimize habitat disruption.

Community engagement: Local communities should be involved in developing communication solutions that meet their needs while respecting their environment. This can help ensure that technological advancements benefit local populations without jeopardizing environmental sustainability.

Climate action: Addressing climate change through aggressive global strategies is essential. Transitioning to renewable energy, reducing carbon emissions, and promoting sustainable practices are vital to prevent further disruptions in these delicate ecosystems.

Life in harsh environments has always been a precarious balancing act. However, the growing influence of extreme weather and disruptive communication technologies threatens to tip the scales. By working together, embracing sustainability, and mitigating the effects of climate change, we can ensure that these regions remain vibrant ecosystems for generations to come.

ABNORMAL IMBALANCE DUE TO RADIATION

The uneasy dance: abnormal imbalance due to radiation and the price of progress

Radiation, the enigmatic force that permeates our universe, has been a constant companion to life on Earth. From the faint hum of cosmic microwave background radiation to the life-sustaining warmth of the sun, it shapes our environment in profound ways. However, human activities have upset this delicate balance, introducing abnormal imbalances due to radiation that threatens the very essence of life—longevity and healthy existence.

Life in the radiation bathtub: a history of coexistence

Life arose and continues to exist amidst a constant barrage of natural radiation sources. Cosmic rays, solar flares, and naturally occurring radioactive materials in the Earth's crust bombard living organisms with ionizing and non-ionizing radiation. This constant exposure has driven the evolution of sophisticated repair mechanisms within cells.

- *DNA repair mechanisms*: Our DNA, the blueprint for life, is particularly vulnerable to ionizing radiation. Organisms have evolved intricate enzymatic pathways to repair DNA damage caused by radiation-induced free radicals and mutations. These repair mechanisms play a vital role in maintaining genomic integrity and preventing the development of cancers and other radiation-induced diseases.
- *Adaptations at the cellular level*: Certain extremophiles, organisms that thrive in extreme environments, showcase remarkable adaptations to radiation. For example, some bacteria possess highly efficient DNA repair systems and antioxidant defenses that allow them to survive in environments with high levels of natural radiation.

The double-edged sword of progress: when greed disrupts the balance

While humans have always been exposed to natural radiation, our thirst for progress has introduced new and potentially harmful sources of radiation.

- *Nuclear power*: The advent of nuclear power generation has provided a significant source of clean energy. However, accidents like Chernobyl and Fukushima highlight the catastrophic consequences of radiation releases, causing widespread contamination and long-term health effects like cancer and genetic mutations.
- *Medical applications*: Radiation therapy plays a crucial role in cancer treatment, utilizing ionizing radiation to target and destroy cancerous cells. However, improper dosage or prolonged exposure can damage healthy tissues and lead to secondary cancers.
- *Nuclear weapons*: The development and testing of nuclear weapons pose a significant threat to life on Earth. Nuclear detonations release immense amounts of radiation, causing immediate casualties and long-term health risks for generations to come. Additionally, the threat of nuclear war and the associated radioactive fallout poses an existential threat to the very fabric of life.

Technical consequences: a cellular and organismic perspective

The abnormal imbalances caused by excessive radiation exposure have profound technical consequences at the cellular and organismic level:

- *Cellular damage*: Ionizing radiation can directly damage DNA, leading to mutations, cell death, and disruptions in cellular processes like cell division and growth. This disrupts tissue homeostasis and can ultimately lead to the development of cancers.
- *Genotoxic effects*: Radiation can induce mutations in germline cells (sperm and egg cells), affecting the genetic makeup of future generations. This can lead to increased susceptibility to diseases, birth defects, and other genetic disorders.
- *Lifespan reduction*: Radiation exposure can shorten lifespan by accelerating aging processes and increasing the risk of age-related diseases like cardiovascular disease and neurodegenerative disorders. This effect is likely due to the accumulation of DNA damage and cellular dysfunction over time.
- *Reproductive issues*: High doses of radiation can impair reproductive health, leading to infertility, miscarriage, and birth defects.

The toll on wildlife: a silent suffering

Animals are not immune to the detrimental effects of radiation imbalances. Wildlife populations near nuclear power plants, waste disposal sites, and areas affected by nuclear accidents can suffer from increased rates of cancer, genetic mutations, and population decline. Additionally, the disruption of ecosystems due to radioactive contamination can have cascading effects on entire food webs.

The current situation necessitates immediate action to mitigate the risks associated with abnormal radiation imbalances.

- *Stricter regulations*: Implementing stricter regulations on nuclear power generation, waste disposal, and medical applications of radiation is crucial. This includes robust safety protocols and regular inspections to prevent accidents and environmental contamination.
- *Investment in renewable energy*: Transitioning to cleaner energy sources like solar and wind power is essential for reducing our reliance on potentially hazardous nuclear energy.
- *Public awareness*: Educating the public about the risks associated with radiation exposure and the importance of radiation safety practices is imperative.
- *Advancements in radiation protection*: Continued research and development in radiation shielding materials, dosimetry techniques, and medical treatments for radiation exposure can help minimize the harmful effects of radiation.

The technical consequences of radiation exposure are evident at the cellular and organismic level, with DNA damage, increased risk of cancers, and reduced lifespan impacting both humans and animals. Wildlife populations near radiation sources suffer silently, with diminished numbers and ecosystem disruption adding to the devastating picture.

The choice we face is stark. We can continue down the path of greed and overexploitation, jeopardizing the well-being of future generations. Or, we can choose to act responsibly. By implementing stricter regulations, investing in renewable energy, fostering public awareness, and continuing research in radiation protection, we can mitigate the risks and restore a semblance of balance.

The future of life on Earth hinges on our ability to manage radiation responsibly. Let us learn from the mistakes of the past and strive for a future where technology and progress co-exist in harmony with the natural world. This is not just a technical challenge but a moral imperative. We stand at a crossroads, and the path we choose will determine the fate of countless species, including our own.

ILL EFFECTS ON HUMAN HABITAT

The Earth, a vibrant tapestry of ecosystems, has provided humanity with a nurturing home for millennia. However, our ever-growing population and insatiable desire for progress have cast a long shadow, threatening the very foundation of this habitat. The detrimental effects of human actions on our living conditions, food choices, and genetic diversity are numerous. These combined forces present a significant challenge for future generations, jeopardizing their well-being and potentially leading to a "larger defeat."

Living conditions: a shrinking paradise

- *Urbanization and pollution*: The rapid growth of urban centers has resulted in a sprawl of concrete jungles, replacing natural habitats with roads, buildings, and factories. This not only destroys vital ecosystems but also leads to air and water pollution, significantly impacting human health. Increased respiratory illnesses, cardiovascular diseases, and waterborne illnesses are just some of the consequences of polluted environments. A 2022 report by the World Health Organization (WHO) estimated that 99% of the global population breathes polluted air.
- *Climate change and natural disasters*: Our reliance on fossil fuels and unsustainable practices have triggered an alarming rise in global temperatures. This disrupts weather patterns, leading to more frequent and intense droughts, floods, heatwaves, and wildfires. These natural disasters displace populations, destroy infrastructure, and threaten

food security, all of which significantly impact living conditions. A 2023 report by the Intergovernmental Panel on Climate Change (IPCC) highlighted the urgency of addressing climate change, emphasizing the increasing frequency and intensity of extreme weather events.

Food choices: a fork in the road

- *Intensive agriculture and monoculture farming*: Our current agricultural practices, focused on maximizing yields, rely heavily on pesticides, fertilizers, and monoculture farming (planting a single crop over a large area). This depletes soil nutrients, reduces biodiversity, and contaminates water sources with agricultural runoff. Additionally, the excessive use of antibiotics in livestock farming contributes to the rise of antibiotic-resistant bacteria, posing a significant public health threat. A 2021 study published in Nature Sustainability highlighted the detrimental effects of intensive agriculture on soil health and biodiversity.
- *Food waste and unsustainable consumption*: The global food system faces a stark reality—a third of all food produced is wasted. This waste occurs throughout the supply chain, from farm to fork. Additionally, our dietary choices, with increasing consumption of processed foods and meat, put further strain on resources and contribute to greenhouse gas emissions. Promoting sustainable food production practices, reducing food waste, and shifting toward plant-based diets are crucial steps toward a more sustainable food system.

Hybridization: a blurred line

- *Loss of genetic diversity*: Modern agricultural practices often rely on high-yielding, hybrid crops. While this increases overall crop yields, it reduces genetic diversity in the agricultural gene pool. This makes crops more susceptible to diseases and pests, potentially leading to crop failures in the face of new pathogens. Additionally, the introduction of invasive species through human activity disrupts native ecosystems and threatens the survival of endemic species.
- *Ethical concerns of genetic modification*: The rise of genetically modified organisms (GMOs) in agriculture raises ethical concerns regarding the potential risks associated with altering the genetic makeup of food crops. The long-term consequences of consuming such modified foods on human health remain largely unknown. Open scientific dialogue and robust regulations are crucial to ensure the safe and ethical development of GMO technology.

The super dominating tendency: a collision course with nature

Our domineering attitude toward nature is perhaps the most significant threat to our habitat. We view ourselves as separate from, rather than part of, the natural world. This perspective fuels our relentless pursuit of exploiting natural resources for immediate gain, with little regard for the long-term consequences.

- *Unsustainable resource consumption*: Our insatiable thirst for resources, from fossil fuels to minerals, is depleting the Earth's finite resources at an alarming rate. This overconsumption disrupts natural resource cycles and jeopardizes the ability of future generations to meet their basic needs.
- *Loss of biodiversity and ecosystem services*: The current mass extinction event, primarily driven by human activities, is depleting the Earth's biodiversity at an unprecedented rate. Every species plays a vital role in its ecosystem, providing essential services like pollination, water purification, and climate regulation. The loss of these species weakens ecosystems, making them less resilient to change and potentially leading to cascading ecological collapses.

The looming threat for future generations: a call to action

The combined impact of these factors paints a bleak picture for future generations. Unsustainable living conditions, a compromised food system, dwindling resources, and a collapsing biosphere threaten not only their well-being but also their very survival.

Transition to renewable energy: Addressing climate change requires a rapid shift away from fossil fuels and toward renewable energy sources like solar, wind, and geothermal power. This transition will not only reduce greenhouse gas emissions but also improve air quality and create new job opportunities.

Sustainable food systems: Promoting sustainable agricultural practices like organic farming, crop rotation, and integrated pest management is crucial. Additionally, reducing food waste and encouraging plant-based diets can significantly contribute to a more sustainable and resilient food system.

Conservation and ecosystem restoration: Protecting endangered species and restoring degraded ecosystems are vital steps toward preserving biodiversity and the vital services ecosystems provide. This includes establishing protected areas, promoting responsible land management practices, and supporting conservation efforts worldwide.

Education and awareness: Educating the public about the environmental challenges we face and fostering a sense of environmental responsibility is crucial. Empowering individuals to make conscious choices in their daily lives, from their consumption habits to their transportation preferences, can have a significant collective impact.

International cooperation: The environmental crisis transcends national borders. International cooperation and collaboration are essential to addressing global challenges like climate change and biodiversity loss. Multilateral agreements, knowledge sharing, and joint initiatives are key to ensuring a sustainable future for all.

The story of humanity's relationship with the Earth can be rewritten. However, it requires a profound shift in perspective—from one of dominance to one of co-existence. We must recognize ourselves as integral parts of the natural world, not separate from it. Sustainable practices, responsible resource consumption, and a deep respect for nature's intricate balance are the cornerstones of a future where both humanity and the environment can thrive. This is not just about protecting some distant future; it's about ensuring that future generations inherit a healthy planet, capable of sustaining life and fostering a vibrant future for all.

REFERENCES

1. Liu, D. (2023). International Energy Agency (IEA). In S. N. Romaniuk & P. N. Marton (Eds.), *The Palgrave encyclopedia of global security studies* (pp. 830–836). Palgrave Macmillan.
2. Nigra, A. E., Chen, Q., Chillrud, S. N., Wang, L., Harvey, D., Mailloux, B., Factor-Litvak, P., & Navas-Acien, A. (2020). Inequalities in public water arsenic concentrations in counties and community water systems across the United States, 2006–2011. *Environmental Health Perspectives*, 128(12). https://doi.org/10.1289/ehp7313
3. Calicioglu, O., Flammini, A., Bracco, S., Bellù, L., & Sims, R. (2019). The future challenges of food and agriculture: an integrated analysis of trends and solutions. *Sustainability*, 11(1), 222. https://doi.org/10.3390/su11010222
4. Schumacher, D. L., Keune, J., Dirmeyer, P., & Miralles, D. G. (2022). Drought self-propagation in drylands due to land–atmosphere feedbacks. *Nature Geoscience*, 15(4), 262–268. https://doi.org/10.1038/s41561-022 -00912-7
5. Ludescher, J., Martin, M., Boers, N., & Schellnhuber, H. J., et al. (2021). Network-based forecasting of climate phenomena. *Proceedings of the National Academy of Sciences*, 118(47), e1922872118. https://doi.org/10 .1073/pnas.1922872118
6. IPCC. (2023). Sections. In Core Writing Team, H. Lee, & J. Romero (Eds.), *Climate Change 2023: Synthesis Report. Contribution of Working Groups I, II and III to the Sixth Assessment Report of the Intergovernmental Panel on Climate Change.* IPCC. https://doi.org/10.59327/IPCC/AR6-9789291691647

7. World Food Programme (WFP). (2023). *WFP Pakistan floods situation report, August 2023.* Situation Report. Posted September 29, 2023. Originally published September 29, 2023

8. Erie, S. P., Kogan, V., & MacKenzie, S. A. (2012). Paradise plundered: fiscal crisis and governance failures in San Diego. *Choice Reviews Online, 49*(7), 49–3969. https://doi.org/10.5860/choice.49–3969

9. NASA Climate Change. (2023, June 15). Arctic sea ice.

10. Climate Change 2021—The Physical Science Basis. (2021). *Chemistry International, 43*(4), 22–23. https://doi.org/10.1515/ci-2021–0407

11. Wang Wang, Y., Hollingsworth, P. M., Zhai, D., et al. (2023). High-resolution maps show that rubber causes substantial deforestation. *Nature, 623*(340), 340–346. https://doi.org/10.1038/s41586-023-06642-z

12. Van Oldenborgh, G. J., Van Der Wiel, K., Sebastian, A., Singh, R., Arrighi, J., Otto, F., Haustein, K., Li, S., Vecchi, G., & Cullen, H. (2017). Attribution of extreme rainfall from Hurricane Harvey, August 2017. *Environmental Research Letters, 12*(12), 124009. https://doi.org/10.1088/1748-9326/aa9ef2

13. Sellitto, P., Belhadji, R., Kloss, C., & Legras, B. (2022). Radiative impacts of the Australian bushfires 2019–2020 – Part 1: large-scale radiative forcing. *Atmospheric Chemistry and Physics, 22*(14), 9299–9311. https://doi.org/10.5194/acp-22-9299-2022

14. Goda, K., Kiyota, T., Pokhrel, R. M., Chiaro, G., Katagiri, T., Sharma, K., & Wilkinson, S. (2015). The 2015 Gorkha Nepal earthquake: insights from earthquake damage survey. *Frontiers in Built Environment, 1.* https://doi.org/10.3389/fbuil.2015.00008

15. Sohail, T. (2023). Committed future ice-shelf melt. *Nature Climate Change, 13,* 1164–1165. https://doi.org/10.1038/s41558-023-01817-y

16. Awolala, D. O., Mutemi, J., Adefisan, E., Antwi-Agyei, P., & Taylor, A. (2022). Profiling user needs for weather and climate information in fostering drought risk preparedness in Central-Southern Nigeria. *Frontiers in Climate, 4.* https://doi.org/10.3389/fclim.2022.787605

17. Ayeb-Karlsson, S., Hoad, A. & Trueba, M. L. (2024). 'My appetite and mind would go': Inuit perceptions of (im)mobility and wellbeing loss under climate change across Inuit Nunangat in the Canadian Arctic. *Humanities and Social Sciences Communications, 11,* 277. https://doi.org/10.1057/s41599-024-02706-1

18. Miara, M. D., Negadi, M., Tabak, S., Bendif, H., Dahmani, W., Ait Hammou, M., Sahnoun, T., Snorek, J., Porcher, V., Reyes-García, V., & Teixidor-Toneu, I. (2022). Climate change impacts can be differentially perceived across time scales: A study among the Tuareg of the Algerian Sahara. *GeoHealth, 6*(11), e2022GH000620. https://doi.org/10.1029/2022GH000620

Chapter 5

Commercial radiation

RACE OF TECHNOLOGY
ADVANCEMENTS IN SPECTRUM

The electromagnetic spectrum, a vast and finite resource, serves as the foundation for all wireless communication technologies. From radio waves to gamma rays, different frequencies within the spectrum cater to diverse applications, ranging from mobile communication networks to satellite navigation and medical imaging. As technological advancements continue to drive the demand for faster data rates, higher capacity, and lower latency, the race for spectrum utilization becomes ever more critical. The proliferation of mobile devices, the rise of the Internet of Things (IoT), and the ever-increasing demand for high-definition video streaming are placing immense strain on existing spectrum allocations. Fifth-generation (5G) mobile networks, with their promise of ultra-fast speeds and ultra-low latency, require access to new and wider frequency bands compared to previous generations. This has led to a shift toward utilizing higher frequency bands, particularly millimeter wave (mmWave) spectrum ranging from 24 GHz to 86 GHz. While mmWave offers significant bandwidth capabilities, its propagation characteristics present challenges due to shorter range and higher susceptibility to signal blockage.

Innovations in antenna design, such as massive MIMO (Multiple-Input and Multiple-Output) systems, are helping to overcome these limitations by utilizing large antenna arrays to focus and steer signals for improved coverage and capacity. Additionally, advancements in beamforming techniques are enabling more efficient use of the mmWave spectrum by directing signals toward specific users rather than broadcasting them omnidirectionally.

Exploiting underutilized spectrum

While exploring new frontiers like mmWave is crucial, another approach involves optimizing the utilization of existing spectrum bands. Cognitive radio technology [1] is emerging as a promising solution in this regard. Cognitive radios are intelligent devices that can dynamically sense and adapt

DOI: 10.1201/9781003590712-5

their transmission parameters based on the real-time availability of spectrum in a specific location. This allows for opportunistic spectrum access, where unused portions of the spectrum can be temporarily utilized by cognitive radio devices [2] without interfering with existing licensed users.

Software-defined radio (SDR) technology plays a vital role in cognitive radio systems. SDRs are programmable radios that can be configured to operate on different frequencies and protocols, making them ideal for dynamic spectrum access. The continued development of SDRs with increased processing power and flexibility will be instrumental in unlocking the full potential of cognitive radio technology.

Spectrum sharing and aggregation

Spectrum sharing mechanisms are being explored to enable diverse applications to coexist within the same frequency band. Licensed Shared Access (LSA) [3] is one such approach, where a portion of a licensed spectrum band is made available for unlicensed use under certain conditions. This allows for increased spectrum utilization while ensuring minimal interference with the primary licensed users.

Spectrum aggregation is another technique that combines multiple fragmented spectrum allocations into a wider, contiguous band. This approach offers significant benefits for applications requiring high bandwidth, such as next-generation mobile networks and wireless backhaul for fixed broadband services. Advancements in signal processing algorithms and network management techniques are crucial for enabling efficient spectrum aggregation and mitigating potential interference issues.

New frontiers: beyond terrestrial communication

The race for spectrum extends beyond terrestrial communication. The increasing demand for global connectivity is driving advancements in satellite communication technologies. Low-Earth Orbit (LEO) satellite constellations are being deployed to provide high-speed internet access to remote and underserved areas. These constellations operate in different frequency bands compared to traditional geostationary satellites, necessitating careful spectrum management strategies to avoid interference between different satellite systems and terrestrial networks.

Furthermore, research and development efforts are underway to explore the potential of utilizing terahertz (THz) frequencies [4] for future wireless communication systems. The THz spectrum offers extremely high bandwidth capabilities, potentially enabling revolutionary applications in areas like ultra-fast data transfer and high-resolution imaging. However, significant challenges exist in developing practical THz systems due to the limitations of current materials and signal propagation characteristics at these frequencies. Spectrum regulators around the world are actively exploring new approaches to manage the ever-growing demand for spectrum. This includes implementing dynamic spectrum allocation mechanisms, facilitating spectrum-sharing agreements between different stakeholders, and streamlining licensing procedures to encourage innovation and investment in new technologies.

Challenges and considerations

Despite the exciting advancements in spectrum utilization, several challenges remain. Efficient spectrum management requires balancing the needs of diverse stakeholders, including mobile network operators, broadcasters, satellite communication providers, and emerging technologies like autonomous vehicles. Furthermore, ensuring fair access to spectrum and preventing harmful interference between different users is critical to maintaining a healthy wireless ecosystem.

The security of wireless communication systems is another crucial consideration. As the spectrum becomes more crowded and transmission techniques become more complex, vulnerabilities to cyberattacks and signal jamming become a growing concern. Developing robust security protocols and implementing effective mitigation strategies will be essential for safeguarding the integrity and reliability of future wireless communication systems. The race for spectrum is a continuous process driven by the relentless pursuit of faster, more reliable, and more ubiquitous wireless communication. Technological advancements are pushing the boundaries of spectrum utilization, opening doors to innovative applications and transforming the way we connect with the world around us. Optimizing existing spectrum allocations, exploring new frequency bands, and fostering a flexible and responsive regulatory environment will be crucial factors in ensuring that the spectrum remains a valuable resource for generations to come.

The future of the race for spectrum promises to be a fascinating one, with ongoing research and development efforts holding the potential to unlock even greater possibilities. From cognitive radio networks that can dynamically adapt to their surroundings to terahertz-based communication systems offering unimaginable data rates, the spectrum landscape continues to evolve at a rapid pace. As technology advances, the race for spectrum will undoubtedly continue, shaping the future of wireless communication and driving innovation across diverse sectors.

FASTER DATA, FASTER LIFE- AUTOMIZATION EVERYWHERE

The double-edged sword of automation

The relentless pursuit of faster data speeds and ubiquitous automation is undeniably transforming our lives. Communication gadgets have become extensions of ourselves, connecting us instantly and offering a seemingly endless stream of information and entertainment. However, this constant connectivity comes at a cost, impacting our data storage needs, pushing us toward a faster-paced lifestyle, and potentially affecting our emotional well-being.

Speed demons: a spectrum of communication gadgets

From smartphones boasting lightning-fast internet connections to smart-watches that track our every step, the pace of technological advancement is dizzying. 4G and 5G networks offer blazing-fast download and upload speeds, enabling real-time video conferencing, seamless streaming of high-definition content, and near-instantaneous file sharing. These advancements enhance our ability to connect and collaborate, fostering a sense of global community and access to information. However, the exponential growth of data consumption [5] accompanying faster speeds creates its own set of challenges. Sharing high-resolution photos and videos, downloading large software updates, and using cloud-based applications all contribute to the ever-increasing demand for data storage. Smartphones may offer ample storage initially, but users often find themselves struggling to keep up with the ever-expanding digital footprint.

From smartphones boasting lightning-fast internet connections to smart-watches that track our every step, the pace of technological advancement is dizzying. Let's delve into specific gadgets and their data speeds:

- *Smartphones*: The workhorses of modern communication, smart-phones like the iPhone 13 Pro Max or the Samsung Galaxy S22 Ultra offer download speeds exceeding up to Gbps on 5G networks. This allows for near-instantaneous downloads of large files, high-definition video streaming without buffering, and real-time video calls. However, such speeds come at a cost, with heavy data usage quickly depleting data plans.
- *Smartwatches*: These wearable devices, like the Apple Watch Series 8 or the Samsung Galaxy Watch 5, typically rely on Bluetooth or Wi-Fi for data transfer. While speeds are slower than smartphones, ranging from 10 Mbps to 100 Mbps, they are sufficient for receiving notifications, tracking fitness data, and streaming music. The limited data storage on smartwatches (often less than 32 GB) necessitates syncing them with smartphones for extensive data management.
- *Laptops*: Modern laptops like the MacBook Pro M2 or the Dell XPS 13 Plus can achieve download speeds exceeding 500 Mbps on Wi-Fi 6 networks. This enables efficient video conferencing, cloud storage access, and large software downloads. However, data storage capacity

varies depending on the model, ranging from 256 GB to 2 TB, requiring users to manage their storage space based on their needs.

- *Smart home devices*: These connected devices, such as Amazon Echo or Google Nest Hub, typically rely on Wi-Fi for data transfer, with speeds varying depending on the model and network. While data usage is generally low for basic functionalities, streaming music or video content can quickly consume data.

The automation paradox: convenience vs. control

The rise of automation is weaving itself into the fabric of our daily lives. Smart homes adjust lighting and temperature based on our preferences, while self-driving cars promise to revolutionize transportation. These advancements offer undeniable convenience, freeing up our time and mental bandwidth for other pursuits. However, the pervasive nature of automation can also lead to a sense of losing control and a disconnect from the tasks we perform. Furthermore, the constant barrage of notifications and alerts bombarding us from our communication devices can contribute to feeling overwhelmed and hinder our ability to focus. The pressure to keep pace with the "always-on" mentality fostered by faster data speeds can be detrimental to our mental well-being, leading to increased stress and anxiety.

The emotional toll: a call for balance

While faster data and automation offer undeniable benefits, it's crucial to acknowledge the potential downsides. Finding a healthy balance between harnessing the power of technology and prioritizing our emotional well-being is paramount. Here are a few steps we can take:

- *Digital detoxification*: Setting aside dedicated time to disconnect from devices and reconnect with ourselves can be immensely beneficial.
- *Conscious consumption*: Being mindful of what data we consume and for what purpose allows us to avoid information overload and focus on what truly matters.
- *Prioritizing physical activity*: Engaging in regular physical activity is a proven way to reduce stress and improve focus, countering the negative effects of a sedentary, fast-paced lifestyle.

The future of communication technology promises even faster data speeds and further integration of automation into our lives. While these advancements hold immense potential, it's vital to approach them with a critical eye, prioritizing a balanced relationship with technology that fosters connection, convenience, and well-being.

DESENSITIVITY IN APPROACH

The desensitized rush: when speed trumps empathy in the modern age

The human experience is a tapestry woven with an intricate dance between progress and compassion. Yet, in the whirlwind of the modern age, a concerning trend emerges—a growing desensitization in our approach to life. This desensitization is fueled by an insatiable hunger for speed, a relentless pursuit of material gain, and a distorted definition of success. Gone are the days of savoring the journey; our focus has shifted to reaching the destination as quickly as possible. Smartphones connect us instantly, yet meaningful conversations dwindle. Fast food satiates our hunger pangs momentarily but leaves us yearning for genuine nourishment. This relentless pursuit of speed spills over into our interactions with others, leaving a trail of emotional neglect in its wake.

The tyranny of "more": a breeding ground for apathy

Our minds, bombarded with a constant barrage of information and stimuli, have become conditioned to crave "more." Social media platforms flaunt curated realities, fostering a comparison trap that breeds feelings of inadequacy. The relentless pursuit of acquiring possessions and achieving a certain financial status becomes the sole measure of worth, casting a shadow on the value of genuine connection and empathy. This obsession with "more" translates into a disregard for the struggles of others. We become desensitized to the plight of those less fortunate, our hearts hardened by the constant barrage of negativity and suffering that floods our news feeds. The urgency to fulfill our own desires eclipses the capacity to truly understand and respond to the needs of those around us.

The erosion of empathy: a vicious cycle

The desensitization we experience isn't merely a societal ill; it has neurobiological implications [6]. Studies have shown that repeated exposure to violence and negativity can desensitize the amygdala, the part of the brain responsible for processing emotions like fear and empathy. This creates a vicious cycle—the more desensitized we become, the less we are affected by the suffering of others, leading to a further erosion of empathy.

The consequences of this desensitization are far-reaching. It weakens the social fabric, fostering a culture of individualism and self-interest. It erodes trust and cooperation, making it difficult to address global challenges that require collective action. It creates a society where emotional detachment becomes the norm, leading to loneliness, isolation, and a decline in mental well-being.

Nature's reckoning: a call to reclaim our humanity

Nature, in its intricate balance, serves as a stark reminder of the interconnectedness of all things. The rampant consumerism [7] and disregard for the environment that accompany our desensitized approach to life are ultimately unsustainable. Climate change, resource depletion, and biodiversity loss serve as stark warnings that a reckoning is upon us if we fail to change course.

Reclaiming our humanity: a path forward

The need of the hour is a conscious effort to decondition ourselves from the relentless pursuit of speed and material wealth. Here are some steps we can take to reclaim our humanity and foster a more compassionate approach to life:

- *Practice gratitude*: Cultivate an attitude of gratitude for the blessings in our life, no matter how small. This shifts the focus away from what we lack and helps us appreciate the simple joys of life.
- *Disconnect to reconnect*: Regularly disconnect from technology to make space for real-world connections. Engage in face-to-face interactions, build meaningful relationships, and invest time in fostering emotional intimacy.
- *Practice mindfulness*: Mindfulness practices like meditation can help us become more aware of our thoughts and emotions. They allow us to cultivate compassion not just for ourselves but also for others.
- *Engage in acts of service*: Volunteering our time and resources for causes we believe in allows us to connect with the needs of others and contribute to a better world.

By adopting a slower, more mindful approach to life, we can cultivate empathy, reconnect with our values, and build a more compassionate society. This isn't just a call for personal transformation; it's a call for collective action. By actively choosing to prioritize human connection, environmental sensitivity, and genuine well-being over the fleeting allure of speed and material wealth, we can create a future where humanity thrives in harmony with nature and each other.

Remember, the human spirit is resilient and capable of immense compassion. By choosing to be more present, fostering deeper connections, and prioritizing empathy, we can rewrite the narrative of the modern age, ensuring that technology serves humanity and not the other way around. Let's choose a path of conscious living, ensuring that the tapestry of human experience continues to be woven with threads of compassion, empathy, and a deep appreciation for the beauty and fragility of life on this planet.

DATA STORAGE CONSTRAINTS

The drowning data dilemma: storage constraints and environmental impacts

The digital age is awash in data. From the trillions of emails sent daily to the constant stream of social media posts and sensor data, the amount of information we generate is exploding. This data deluge has created a major challenge: data storage constraints [8]. Storing this ever-increasing volume of information efficiently and sustainably requires innovative solutions and a critical examination of current practices.

Demystifying data storage: techniques and technologies

Data storage relies on a variety of technologies, each with its own strengths and limitations. Here are some key techniques:

- *Hard disk drives (HDDs)*: These traditional workhorses use magnetic platters to store data. They are relatively inexpensive and offer high capacity but are slower and more susceptible to physical damage compared to newer technologies.
- *Solid-state drives (SSDs)*: SSDs utilize flash memory chips to store data electronically, allowing for faster read/write speeds and lower latency. However, they typically come with higher costs per gigabyte and limited storage capacities compared to HDDs.
- *Tape storage*: Magnetic tape remains a cost-effective option for long-term archival storage. While slow to access, tape offers good durability and high density, making it ideal for infrequently accessed data.
- *Cloud storage*: Cloud storage solutions offer remote data storage accessed over the internet. This provides scalability and flexibility but raises concerns about data security and dependence on internet connectivity.

The choice of storage technique depends on factors such as cost, performance requirements, data access frequency, and security considerations.

Data centers: the power-hungry guardians of the digital world

Data centers, sprawling facilities housing massive server banks, form the backbone of the digital infrastructure. They store and process the vast quantities of data generated globally. However, this essential function comes at a significant environmental cost. Data centers [9] are notorious for their energy consumption. Cooling the servers to prevent overheating requires immense amounts of electricity. A study by the Natural Resources Defense Council (NRDC) revealed that data centers in the USA alone consume roughly 2% of the nation's total electricity usage, equivalent to the annual output of 50 power plants.

To meet this demand, data centers often rely on fossil fuels, contributing to greenhouse gas emissions and climate change. Furthermore, the constant water usage for cooling systems can be a major concern in water-scarce regions.

Real-world example: In 2020, Apple announced plans to build a massive data center in North Carolina. While touted for job creation, the project faced criticism for its potential strain on the state's already stressed water resources, particularly during dry seasons.

Marine life caught in the digital current

The environmental impact of data centers extends beyond energy consumption and water usage. A lesser-known concern is the impact on marine life. A growing trend involves building data centers near bodies of water, as seawater offers a readily available and seemingly limitless coolant. However, this practice has detrimental consequences for marine ecosystems.

The process of drawing in and discharging large volumes of seawater can harm delicate marine life. Intake pipes can trap and injure fish and other organisms. Additionally, the discharge of warm water back into the ocean disrupts thermal equilibrium, harming coral reefs and other temperature-sensitive ecosystems.

Real-world example: In 2019, a data center in Singapore was linked to a mass die-off of sea life. Investigations revealed that the facility's intake pipes were drawing in large quantities of small marine organisms, leading to their demise.

The distilled water dilemma: a paradox of scarcity

One surprising aspect of data center operation involves the use of distilled water. While the world faces a growing water crisis, data centers often utilize distilled water for cooling purposes. This seemingly wasteful practice stems from the need for highly purified water to prevent corrosion within the sensitive cooling systems. However, the energy-intensive process of distillation adds another layer to the environmental impact of data centers. In water-scarce regions, using precious freshwater for non-essential purposes creates a conflict with the needs of local communities.

Real-world example: In 2018, a report highlighted the case of a data center in Chennai, India, a city grappling with water shortages. The facility's reliance on distilled water raised concerns about prioritizing the needs of a private company over the well-being of the city's residents.

Moving forward: toward sustainable data storage

- *Data optimization and archiving*: Implementing data lifecycle management practices can help reduce storage demands. This includes deleting obsolete data, migrating infrequently accessed data to cost-effective archival solutions like tape storage, and optimizing data compression techniques.
- *Circular economy for hardware*: Extending the lifespan of data storage hardware through refurbishing and responsible end-of-life recycling can minimize resource depletion and electronic waste generation.
- *Consumer awareness and responsible data management*: Educating consumers about the environmental impact of their digital footprint can encourage responsible data management practices. This includes minimizing unnecessary data generation, adopting cloud storage providers with sustainability commitments, and regularly cleaning up personal data.

The data storage conundrum demands a collaborative effort from individuals, businesses, and policymakers. Embracing innovative technologies, prioritizing sustainability, and fostering responsible data practices are crucial steps toward building a future where the digital world and the natural world can coexist in harmony. The choices we make today will determine whether data continues to be a resource or becomes a burden. By embracing a future where efficiency, sustainability, and responsible data management are core principles, we can ensure that the information age is not synonymous with an environmental crisis. The power lies in our hands—individuals, corporations, and governments—to navigate the data deluge responsibly and build a digital future that thrives alongside a healthy planet.

Empowering users: tools and techniques for data management

The fight against data storage constraints extends beyond large-scale data centers. Individuals can make a significant impact by adopting responsible data management practices. Here are some tools and techniques to empower users:

- *Duplicate file finders*: These software programs scan storage devices to identify and eliminate redundant files. This can be particularly helpful for photos, where multiple versions or different edits may be saved unknowingly.
- *Cloud storage management apps*: Many cloud storage providers offer apps to manage files across different devices. These apps allow users to easily organize, share, and back up data, optimizing storage utilization.

- *Automatic backup and sync tools*: Services like Google Photos or iCloud Photos can automatically back up photos and videos to the cloud, freeing up space on devices. Users can then choose to delete local copies once the backup is complete.
- *Data archiving solutions*: For infrequently accessed files with sentimental value, consider archiving them on external hard drives or dedicated archival storage services. This frees up cloud storage space for more frequently used data.

The psychology of data hoarding

Data hoarding [10] can stem from various psychological factors, including:

- *Fear of loss*: The desire to hold on to everything, driven by the fear of losing precious memories or important information.
- *Decision paralysis*: The inability to decide what to delete due to the emotional attachment associated with digital files.
- *Perfectionism*: The need to maintain a complete and unedited digital record, even if it means storing unnecessary duplicates.

Decluttering the digital landscape: benefits abound

While the initial process may require effort, decluttering your digital space offers several benefits:

- *Reduced storage costs*: By eliminating unnecessary data, you may be able to downsize your cloud storage plan, leading to cost savings.
- *Improved device performance*: With less data to manage, devices can run faster and experience fewer storage-related slowdowns.
- *Enhanced security*: A cluttered digital space with numerous files increases the risk of data breaches or malware infections. Decluttering minimizes these risks.
- *Peace of mind*: Having a well-organized digital space can lead to a sense of calm and control, knowing that important information is readily accessible.

By adopting these tools and techniques, individuals can become active participants in the fight against data storage constraints. Taking control of your digital footprint empowers you to not only contribute to a more sustainable data ecosystem but also experience the personal benefits of a clutter-free digital life.

ENERGY CUTOFFS

The looming energy crisis: a call for conservation and storage

The specter of energy cutoffs, once a distant concern, is rapidly becoming a stark reality. As global populations surge and industrialization accelerates, the strain on our energy resources is intensifying. This escalating crisis necessitates a paradigm shift in our approach to energy consumption and management. The key to mitigating the impending energy crunch lies in judicious utilization, coupled with robust storage solutions.

Energy, the lifeblood of modern civilization, is finite. While renewable sources are gaining traction, they are yet to fully supplant conventional fuels. The intermittent nature of solar and wind energy exacerbates the challenge, underscoring the critical need for efficient storage technologies. Moreover, the burgeoning electric vehicle market places an additional strain on the grid, necessitating smarter charging strategies and expanded infrastructure.

To avert energy catastrophes, we must prioritize conservation. This entails a multi-pronged approach, encompassing both individual and systemic changes. At the individual level, adopting energy-efficient appliances, optimizing home insulation, and cultivating energy-conscious habits can yield substantial savings. For industries and governments, the focus should be on technological advancements, such as improving energy efficiency in manufacturing processes and promoting sustainable transportation modes.

Simultaneously, investing in energy storage is imperative. Batteries, both large-scale and portable, hold immense promise. Advancements in lithium-ion technology have paved the way for more efficient and affordable storage solutions. However, research and development must continue to explore alternative battery chemistries and emerging technologies like flow batteries and solid-state batteries. Beyond batteries, other storage options such as pumped hydro storage, compressed air energy storage, and thermal energy storage deserve attention.

The grid itself requires a makeover to accommodate the influx of renewable energy and the growing demand for electricity. Smart grids, equipped with advanced metering and communication technologies, can optimize energy distribution, reduce losses, and facilitate the integration of renewable sources. Additionally, microgrids can enhance grid resilience by enabling localized power generation and distribution.

Energy efficiency is not merely about reducing consumption but also about maximizing output. Cogeneration, or combined heat and power, is a prime example of efficient energy utilization. By capturing waste heat from power generation processes, cogeneration significantly improves overall energy efficiency. Furthermore, industrial symbiosis, where waste products from one industry serve as raw materials for another, can create closed-loop energy systems, minimizing waste and resource depletion.

Behavioral changes are equally crucial. Raising awareness about energy consumption patterns and promoting energy-efficient lifestyles is essential. Education and public campaigns can empower individuals to make informed choices and reduce their energy footprint. Governments can incentivize energy-saving behaviors through policies and regulations, such as feed-in tariffs for renewable energy producers and energy efficiency standards for appliances.

The transition to a sustainable energy future is a complex and multifaceted challenge. It demands collaboration between governments, industries, and individuals. By embracing energy conservation, investing in storage technologies, and modernizing the grid, we can mitigate the risks of energy cutoffs and build a more resilient and sustainable energy system for generations to come. It is imperative to recognize that the choices we make today will shape the energy landscape of tomorrow.

REFERENCES

1. Ahmad, W. S. H. M. W., Radzi, N. a. M., Samidi, F. S., Ismail, A., Abdullah, F., Jamaludin, M. Z., & Zakaria, M. N. (2020). 5G technology: Towards dynamic spectrum sharing using cognitive radio networks. *IEEE Access, 8,* 14460–14488. https://doi.org/10.1109/access.2020.2966271

2. Haykin, S. (2005). *Cognitive radio: Brain-inspired wireless communications.* John Wiley & Sons.

3. Matinmikko, M., Okkonen, H., Palola, M., Yrjola, S., Ahokangas, P., & Mustonen, M. (2014). Spectrum sharing using licensed shared access: the concept and its workflow for LTE-advanced networks. *IEEE Wireless Communications, 21*(2), 72–79. https://doi.org/10.1109/mwc.2014.6812294

4. Zhang, Z., Xiao, Y., Ma, Z., Xiao, M., Ding, Z., Lei, X., Karagiannidis, G. K., & Fan, P. (2019). 6G wireless networks: vision, requirements, architecture, and key technologies. *IEEE Vehicular Technology Magazine, 14*(3), 28–41. https://doi.org/10.1109/mvt.2019.2921208

5. Mao, Y., You, C., Zhang, J., Huang, K., & Letaief, K. B. (2017). A survey on mobile edge computing: the communication perspective. *IEEE Communications Surveys & Tutorials, 19*(4), 2322–2358. https://doi.org/10.1109/comst.2017.2745201

6. Poddar, S., Kar, B., Samanta, S., & Bhaumik, A. (2021). *Human resource management and mental health-a psychosocial aspect.* Lincoln Research and Publications Limited, Australia in Collaboration with Lincoln University College. https://doi.org/10.46977/book.2021.hrmmh

7. Webb, J. W., Ireland, R. D., & Ketchen, D. J. (2014). Toward a greater understanding of entrepreneurship and strategy in the informal economy. *Strategic Entrepreneurship Journal, 8*(1), 1–15. https://doi.org/10.1002/sej.1176

8. Hou, Y., Huang, J., Xie, D., & Zhou, W. (2022). The limits to growth in the data economy: how data storage constraint threats. *SSRN Electronic Journal*. https://doi.org/10.2139/ssrn.4099544

9. Kandula, S., Sengupta, S., Greenberg, A., Patel, P., & Chaiken, R. (2009). The nature of data center traffic. In *Proceedings of the 9th ACM SIGCOMM conference on Internet measurement (IMC '09)* (pp. 202–208). Association for Computing Machinery. https://doi.org/10.1145/1644893.1644918

10. Gormley, C. J., & Gormley, S. J. (2012). Data hoarding and information clutter: the impact on cost, life span of data, effectiveness, sharing, productivity, and knowledge management culture. *Issues in Information Systems*, *13*(2), 90–95. https://doi.org/10.48009/2_iis_2012_90-95

Chapter 6

Living better

WISE CHOICE IN AUTOMATION OF LIFE

The hidden price of comfort

As time has progressed without stopping, our dependence on technology and automation to simplify our lives has grown. We are surrounded by tools and systems that are meant to make our lives easier and more efficient from the moment we wake up until we go to sleep. Undoubtedly, these developments have had many positive effects, but they have also come at a high cost that is sometimes hidden and unrecognized. Our obsession with automation has caused us to overlook important facets of the human experience. Real relationships have been substituted for virtual ones, physical activity for sedentary lifestyles, and thoughtful consumption for impulsive buys. Technology's ease of use has turned into a double-edged sword, drawing us into a loop of dependency that is steadily harming both our planet's health and our own well-being.

Linking back to our origins

We need to take a voyage of rediscovery and go back to the basic ideas that have supported human civilization for thousands of years in order to overcome this rising issue. We can cultivate a more positive relationship with ourselves, our communities, and the natural world by adopting a more straightforward and grounded lifestyle.

Living does not necessarily mean suffering or deprivation. Rather, it entails making deliberate decisions about our time management, social contacts, and purchasing habits. We can live more meaningful and full lives while leaving less of an ecological footprint if we value connections over isolation, quality over quantity, and experiences over stuff.

Among the many things we may do to enhance our sense of well-being and re-establish a connection with nature are gardening, cooking, and manual labor. A more balanced and contented life can also be attained by

DOI: 10.1201/9781003590712-6

engaging in community events, building relationships, and spending time outside.

Advantages of living a simpler life

Making the transition to a simpler lifestyle has many benefits for people and society at large. We can conserve resources, slow down climate change, and enhance the quality of the air and water by lowering our dependency on material items and energy-intensive activities. A slower pace of life can also improve mental clarity, encourage creativity, and lessen stress.

Additionally, leading a simple life helps foster social cohesiveness and builds our communities. Through collective endeavors and patronizing neighborhood companies, we can build a stronger, more cohesive community.

Striking a balance

It's crucial to remember that giving up automation and technology entirely is neither a desirable nor a realistic option. Many of these developments have resulted in notable enhancements in the fields of communication, education, and healthcare. The secret is striking a balance between utilizing technology and protecting ecological health and human connection. We can profit from both worlds if we incorporate aspects of simpler living into our everyday activities. For instance, we can spend quality time with people who are close by and stay in touch with loved ones who are far away, thanks to technology. Automation can help us focus on awareness and being present while streamlining duties.

An appeal for intervention

We need to come together and accept a paradigm change if we want to build a more sustainable and satisfying future. This entails opposing the dominant consumerist culture, promoting laws that advance environmental preservation and well-being, and standing up for companies that value moral conduct.

People can also have a significant impact by making deliberate decisions in their daily lives. Through waste reduction, conscious consumption, and emphasizing experiences over material goods, we may have a positive impact toward a more equal and sustainable world.

The final say in the matter is ours. We have the option to build a future where human welfare and the health of the earth are given priority, or we can continue to blindly seek a life of limitless consumerism and technology dependence. The benefits of living a simpler, more contented life are incalculable, even though the journey may initially appear difficult.

The effects of consumerism on the environment

- *Depletion of resources*: Pollution, climate change, and resource depletion are all major effects of consumerism on the environment. For instance, the manufacturing of commodities frequently necessitates the use of non-renewable resources, such as fossil fuels, which contaminate the air and water. Emissions of greenhouse gases can also result from the movement of commodities.
- *Pollution*: There are several ways in which consumerism adds to pollution. For instance, hazardous waste produced during the creation of items can contaminate the air, water, and soil. Pollution can also result from the disposal of consumer products like plastics and electronics.
- *Climate change*: One of the main causes of climate change is consumerism. One major source of greenhouse gas emissions is the manufacture and transportation of products. Climate change is further exacerbated by the use of energy-intensive products like air conditioners and automobiles. Consumption has far-reaching and detrimental effects on the ecosystem that affect the entire globe. We can, however, take certain actions to lessen our influence. We can decide to buy fewer things, buy used items, and recycle as much as we can. Additionally, we can help companies that are dedicated to sustainability. We may contribute to preserving the environment for coming generations by implementing modest adjustments in our own lives.

FREQUENCY CHAOS

Chaos in frequency: the squeeze of spectrum

The electromagnetic spectrum is similar to a global commons since it is a finite resource. Our contemporary world is built on this shared resource, which powers everything from satellite navigation to broadcasting and telecommunications [1]. However, there is a never-before-seen strain on this essential resource. A perfect storm of frequency congestion is being created by the growing number of wireless devices, the unquenchable want for data, and the introduction of new technology.

The scramble of spectrum

The spectrum, which was formerly thought to be very wide, is getting closer together. Every emerging technology, including IoT and 5G, wants a piece of the action. There is competition for a finite resource among mobile providers, broadcasters, government organizations, and private businesses. This has caused a mad scramble for spectrum, with countries and companies fighting for every megahertz that is available. The ineffective use of the available spectrum exacerbates the issue. A great deal of spectrum is either unused or assigned to services that may use less valuable bandwidth. Furthermore, because technology is developing so quickly, spectrum allocations frequently become out of date, which stifles efficiency and creativity.

Reusing frequency: a sword with two edges

An essential idea in wireless communications is frequency reuse. Geographically separating users enables several users to share the same frequency channel. Although this method has been quite helpful in optimizing spectrum use, there are drawbacks.

Interference is an increasing worry as more and more devices vie for the same frequency. Closely spaced users using the same frequency may cause their signals to collide, resulting in poor performance and missed calls. This is especially troublesome in places with high population density where frequency reuse is limited. Furthermore, it becomes more challenging to efficiently regulate interference due to the growing complexity of wireless networks. Coordination of the diverse wireless environment is made difficult by the spread of tiny cells, Wi-Fi hotspots, and other unlicensed devices.

The cost of taking on too much

The environment is overloaded as a result of the unrelenting demand for spectrum. Network congestion, slower speeds, and a decline in quality of service are the results of networks' inability to handle the constant increase in data traffic. Consequently, this affects the user experience and hinders the creation of new services and applications.

Furthermore, excessive spectrum use may have negative environmental effects. Carbon emissions are a result of the increased power consumption of wireless infrastructure and devices. Concerns over possible health impacts have also been raised by electromagnetic radiation from wireless gadgets, despite the fact that scientists generally agree that the radiation levels from these devices are too low to be harmful.

Choosing a course of action

Taking on the frequency chaos calls for a multifaceted strategy. Optimizing spectrum allocation and utilization requires the involvement of spectrum management authorities. This entails locating unused spectrum, advocating for spectrum sharing, and fostering the creation of new, efficient technology.

To create systems and technologies that are more effective and spectrum-friendly, industry participants must also innovate. This could entail improvements in network optimization, modulation methods, and antenna technology. In order to harmonize spectrum regulations and enable world-wide spectrum sharing, international cooperation is ultimately required. Countries can maximize the use of the global spectrum and prevent duplication of effort by cooperating.

In fact, there is tremendous strain on the frequency spectrum, making it a valuable resource. We need to take a comprehensive approach that includes effective spectrum management, technological innovation, and international cooperation in order to prevent a crisis. We can only guarantee that this essential resource keeps fostering the expansion of our digital economy and society by banding together.

The secret to unlocking frequency potential: spectrum sharing

It is essential to maximize spectrum use due to the increasing demand for it. A viable way to reduce frequency congestion and meet the expanding demands of wireless services is through spectrum sharing.

Spectrum sharing: what is it?

The dynamic distribution of spectrum resources among several users or services is known as spectrum sharing. Sharing, as opposed to conventional fixed spectrum allotment, allows for the efficient and adaptable use of the spectrum, optimizing its capacity and usefulness.

Spectrum sharing types

- *Dynamic spectrum access (DSA)*: Under DSA, devices without licenses can use unclaimed spectrum bands that are reserved for users with licenses. By making use of idle spectrum resources, this method improves spectrum efficiency.
- *Cognitive radio*: These clever gadgets can sense their surrounding spectrum, modify their transmission settings, and share the airwaves with authorized users without interfering.
- *Licensed shared access (LSA)*: LSA permits, subject to certain restrictions and coordination procedures, the sharing of a spectrum band by several licensed users.

Advantages of spectrum sharing

- *Greater efficiency in using the spectrum*: Sharing allows several users to share the same spectrum, which maximizes the use of the scarce resource.

- *Enhanced innovation*: By opening up possibilities for new services and business models, spectrum sharing promotes innovation.
- *Enhanced network performance*: Sharing can improve network coverage and capacity, which will benefit users.
- *Economic growth*: By fostering the development of new industries and jobs, efficient spectrum utilization can stimulate economic growth.

Examples and case studies

- *White spaces*: Spectrum sharing has the ability to close the digital divide, as shown by the use of TV white spaces for broadband access.
- *5G and spectrum sharing*: In order to improve capacity and coverage and enable a broad range of applications, 5G networks make use of spectrum-sharing technology.
- *IoT and spectrum sharing*: In order to facilitate the widespread networking of IoT devices, spectrum sharing is necessary.

Hence, in order to overcome the problems caused by frequency congestion and realize the full potential of the spectrum, spectrum sharing is an effective strategy. Users and service providers can both gain from a more dynamic and efficient wireless ecosystem that can establish by carefully weighing the technical, regulatory, and economic implications of spectrum sharing.

The legislative structure of spectrum sharing

A strong legal foundation [2] is necessary for spectrum sharing to be implemented successfully. It offers the rules, rewards, and enforcement tools required to maintain effective spectrum usage, foster innovation, and balance the interests of various spectrum users.

Important elements of a regulatory framework for spectrum sharing

- *Spectrum allocation and licensing*: Policies for spectrum allocation must be transparent and adaptable. Depending on the particular frequency band and use case, licensing models might vary from shared access licenses to exclusive licenses.
- *Spectrum management and monitoring*: Proper spectrum management is keeping an eye on how the spectrum is being used, spotting interference problems, and acting accordingly.
- *Interference management*: In shared spectrum situations, safeguarding licensed users and preventing harmful interference require strong interference management systems.

- *Dispute resolution*: To guarantee equitable and effective spectrum use, there must be clear processes in place for addressing disagreements among spectrum users.
- *Incentive mechanisms*: Auctioning off spectrum rights or offering tax benefits for investments related to sharing are examples of regulatory incentives that can promote spectrum sharing.
- *Data security and privacy*: It's critical to safeguard user information and provide safe communication in shared spectrum networks.

Case studies

- *FCC's spectrum-sharing initiatives*: With programs like TV white spaces and licensed shared access, the Federal Communications Commission (FCC) in the USA has led the way in spectrum sharing.
- *European Union's spectrum policy*: Harmonization and spectrum efficiency are the main goals of the EU's coordinated approach to spectrum management.

Spectrum sharing in agriculture: a partnership that increases yield

Agriculture is going through a digital revolution, with technology becoming more and more important in raising food security, sustainability, and production. Because spectrum sharing makes a vast array of agricultural uses possible, it has the potential to greatly assist in this transformation.

Accurate farming

- *Real-time data collection*: By using sensors, unmanned aerial vehicles (UAVs) can gather information on pest infestations, soil moisture, and crop health.
- *Remote sensing*: Spectrum sharing can make it easier to employ satellite pictures for yield prediction and large-scale agricultural monitoring.
- *Machine-to-machine communication*: Facilitates the effective sharing of data for improved farming techniques between agricultural machinery and sensors.

Astute agriculture

- *IoT devices*: Spectrum sharing can support the growing number of IoT devices—such as livestock trackers, weather stations, and soil moisture sensors—used in agriculture.

- *Automation*: For accurate navigation and operation, autonomous tractors and harvesters depend on dependable wireless connectivity.
- *Supply chain management*: Reliable wireless communication is necessary for tracking agricultural products from farm to fork efficiently.
- *Rural connectivity*: In order to close the digital divide, spectrum sharing is crucial because many agricultural areas have poor connectivity.
- *Data privacy*: To maintain farmer security and trust, sensitive agricultural data must be protected.
- *Interference management*: One of the main challenges is preventing interference between other spectrum users and agricultural devices.

Case studies

- *Australia's digital agriculture roadmap*: With an emphasis on facilitating UAV operations and Internet of Things applications, the Australian government has recognized spectrum sharing as a critical enabler for digital agriculture.
- *Precision farming initiatives*: To increase crop yields and resource efficiency, a number of nations are testing precision farming initiatives that make use of spectrum sharing.

RESTRICTING EMISSIONS

The sad truth: industrial pollution and global profit before planet

Our earth has been plagued by the enduring impact of the unrelenting quest for economic expansion. As a result of this quest, industrial emissions have grown to represent a serious danger to the environment and the welfare of coming generations. The truth is a sharp contrast to the strict regulations and international agreements that exist: industries frequently put short-term profits ahead of long-term sustainability, which results in a deadly disrespect for the environment.

The industrial imperative: capitalism over environment

There is a lot of pressure to provide steady financial returns. All industries, including manufacturing and energy generation, are mired in a never-ending cycle of profit maximization and cost reduction. This frequently translates into taking shortcuts, such as dodging environmental laws or making minimal investments in pollution control technology [3]. Increased profitability, a competitive edge, and shareholder pleasure are the clear financial rewards. The long-term effects on the environment, however, are disastrous.

Industrial emissions are directly responsible for air pollution, which is brought on by the production of hazardous gases and particulate matter. As a result, there is currently a global health catastrophe, with millions of people dying young from respiratory illnesses. Industrial waste contaminates water bodies, posing a threat to human health as well as aquatic life. Because of greenhouse gas emissions, climate change is wreaking havoc on ecosystems, causing extreme weather, rising sea levels, and a decline in biodiversity.

Regulatory gaps and enforcement difficulties

To reduce industrial emissions, governments all around the world have enacted strict environmental rules. But, because these rules are complicated and the enforcement is lax, there are now chances for industries to take advantage of gaps in the law. Political influence, corruption, and bribery frequently make it difficult for environmental regulations to be implemented effectively. In addition, the sheer size of industrial operations makes compliance and monitoring extremely difficult. Furthermore, rather than being proactive, the regulatory environment is frequently reactive. Rather than foreseeing future issues, environmental regulations are regularly changed in response to emergencies. This reactive strategy frequently gives industries enough time to adjust and come up with workarounds for new laws.

Temporary benefits, prolonged suffering

Many industries have become ignorant of the long-term effects of their actions due to the short-term focus on profit maximization. While the potential advantages of sustainable practices are frequently disregarded, the financial consequences of environmental harm are frequently underestimated. Investing in clean technology, for example, can ultimately result in cost savings as well as enhanced consumer loyalty and brand reputation.

Furthermore, there are significant social and financial costs associated with environmental degradation. Examples include the financial burden on healthcare systems from diseases linked to pollution, the reduction in agricultural output brought on by climate change, and the expenses incurred in recovering from disasters [4]. Even though certain industries profit from their unsustainable methods, society as a whole eventually bears the costs of these activities.

A way ahead: long-term change

The problem of industrial emissions calls for a multifaceted solution. Governments need to bolster environmental laws, step up enforcement, and offer rewards for eco-friendly behavior. Industries need to focus on environmental responsibility and take a long-term view. Customers who support companies that make strong commitments to sustainability have a critical role to play. Making the shift to a low-carbon economy is crucial. Important initiatives include developing clean technologies, encouraging energy efficiency, and investing in renewable energy sources. The concepts of the circular economy can aid in lowering resource usage and waste. In addition, addressing the global aspect of climate change requires international cooperation. The attainment of a sustainable future necessitates a radical reconfiguration of our priorities and values. We have to acknowledge that environmental health cannot be sacrificed for economic progress. Together, we can build a future where planetary well-being and prosperity coexist.

The shadow of the smokestack: industrial emissions and the drive for profit

Once a ray of hope for progress, the industrial revolution has left a lengthy and sinister legacy. Industrial emissions and their unintended consequence have become a global crisis. The unrelenting drive for economic expansion has frequently made short-term profits the priority over sustainability in the long run, disregarding the environment and the welfare of future generations in the process.

The fundamental struggle between profit and the environment is at the core of this problem. Industries frequently find themselves in a difficult situation because of the constant need to generate financial returns. The attraction of quick money can sometimes eclipse the environmental costs over time. This has taken many different forms, from blatant contempt for the law to deceptive tactics that maximize profits at the expense of the environment.

The repercussions are severe. Air pollution is now a major global health concern since it is a direct effect of industrial emissions. Water bodies are having difficulty sustaining life due to contamination from industrial waste. Global ecosystems are being severely impacted by climate change, which is being caused by greenhouse gas emissions. These are not hypothetical dangers; rather, they are urgent issues impacting millions of people. In response, governments have implemented a patchwork of laws aimed at reducing emissions from industry. On the other hand, enforcement is extremely difficult. Effective implementation is frequently hampered by political meddling, corrupt practices, and the sheer magnitude of industrial activities. Furthermore, the regulatory structure itself struggles to keep up with the quick speed of industrial change and is frequently reactive rather than proactive.

Unknowingly, the consumer contributes to this intricate equation. Industrial activity is driven by consumer demand for items, many of which are produced under problematic environmental conditions. Although consumers' awareness of sustainable purchasing is improving, their decisions are still influenced by the appeal of inexpensive, easily accessible items. A change of viewpoint is essential. The environmental costs of an industry's activities must be internalized, and sustainability must be treated as a primary economic strategy rather than an afterthought. Governments must level the playing field so that companies that put sustainability first receive rewards and those that don't receive penalties. Customers need to start being pickier and choose goods with less of an impact on the environment.

Making the shift to a low-carbon economy is imperative, not just a choice. Important initiatives include developing clean technologies, encouraging energy efficiency, and investing in renewable energy. The concepts of the circular economy can aid in lowering resource usage and waste. In order to address the global dimension of climate change, international cooperation is essential.

In the end, the problem is to redefine development. GDP should not be the only metric used to gauge economic progress; social justice, the environment, and long-term sustainability should also be considered. We need to find a solution to this complicated equation if we want to make sure that the earth is habitable for the coming generations. Incremental modifications are no longer appropriate. We must act boldly and decisively. It is essential to our planet's future.

CHOICE OF DATA SPEEDS

The data speed dilemma

An economic, marketing, and technological perfect storm

One of the hallmarks of the digital age is the unrelenting quest for faster internet speeds. Data consumption is no longer a luxury but a need, thanks to the concerted efforts of telecommunication firms, device makers, and content providers. These players have developed a symphony of technological innovations, marketing tactics, and financial incentives. The complex interactions between these variables, looking at how they have fostered a voracious need for speed and ultimately raised consumer costs, are a concern now.

The need for technology

This problem is rooted in the speed at which technology is developing. Moore's Law, which states that integrated circuit transistor counts will double every two years, has sparked an unrelenting quest for processing power. As a result, there is now a greater need for bandwidth to satisfy the ravenous appetite of these increasingly sophisticated gadgets.

Developments in wireless communication technologies, including 3G, 4G, and now 5G, have increased network capacity to process higher data quantities at previously unheard-of rates. Thanks to these technical advances, we now have a positive feedback loop whereby network operators are encouraged to invest in infrastructure upgrades as quicker devices require higher speeds.

The machine for marketing

A consumer culture that is fixated on speed has been cultivated in large part by telecommunications providers. Faster data speeds have been deftly positioned by marketing efforts as a status, convenience, and advancement indicator. Customers are drawn in by the promise of lag-free online gaming, flawless streaming, and quick downloads.

Furthermore, the implementation of tiered data plans has indirectly pushed users toward higher data consumption. These plans incentivize users to increase their data usage in order to offset the higher costs by providing bigger data allowances at premium prices. This has caused the average amount of data consumed by each user to steadily rise, along with aggressive promotional offers and short-term sales.

The ecosystem of devices

The desire for faster speeds has been further compounded by the rise of data-hungry devices. Due to their constantly developing capabilities, smartphones are becoming the main sources of data usage. Smart TVs, tablets, and even smart home appliances add to the total amount of data being used.

By adding features that need fast connectivity, device makers have profited from this trend. Cloud-based services, augmented reality apps, and high-resolution screens have become commonplace, necessitating fast and dependable networks.

It's the right price, but is it? Higher internet speeds are in high demand due to a confluence of factors, including technological breakthroughs, aggressive marketing, and an expanding ecosystem of data-hungry gadgets. Data prices have increased by telecommunication firms in response to the massive costs associated with spectrum acquisition and infrastructure upgrades.

Although they appear to be a solution, the introduction of unlimited data plans has frequently been accompanied by unstated fees. Fair usage guidelines, data throttling, and network congestion have reduced the value of these contracts, leaving customers unhappy.

The path ahead

Because of customer demand and technological advancements, there will probably always be a persistent search for faster internet speeds. But it's crucial to find a balance between affordability and technological advancement.

The real cost of providing high-speed services must be reflected in pricing models that telecommunications companies adopt. Governments can be extremely important by fostering competition, making infrastructure investments, and guaranteeing consumer protection.

In the end, the difficulty is in utilizing technology to enhance people's lives without sacrificing sustainability or affordability. We can guarantee that everyone can take advantage of high-speed internet by promoting a culture of responsible consumption and supporting innovation in data efficiency.

Possible subjects for additional research

- How various socioeconomic and demographic groups are affected by data speed.
- How laws and rules from the government affect how fast data are transmitted.
- The effects of growing data usage and network infrastructure on the environment.
- The possibility of market disruption from other technologies like low-Earth orbit constellations or satellite internet.

We may obtain a more thorough grasp of the intricate elements influencing the need for faster data speeds and the difficulties in fulfilling this demand by going further into these areas.

Data speed's effect on industries

In addition to altering customer behavior on an individual basis, the unrelenting quest for faster data speeds has completely changed entire sectors. High-speed internet access has spurred creativity, upended established company structures, and produced new opportunities. However, unequal access to high-speed internet has also made digital divides worse, which has an effect on social justice and economic progress.

The economic divide and digital divide

There is no denying the link between economic development and internet speed. Strong, fast-paced infrastructure areas frequently draw investment, encourage innovation, and generate employment. Companies in these sectors can take advantage of cutting-edge technologies to boost output, expand into new markets, and enhance client interactions.

On the other hand, places with little or no high-speed internet confront major difficulties. Access to basic services like healthcare and government subsidies is restricted, educational prospects are impeded, and small enterprises face difficulties competing. This digital divide can cause social instability and sustain economic inequality.

Sector shift

- *Healthcare*: High-speed internet has completely changed medical research, telemedicine, and remote patient monitoring. Nevertheless, the reach of these services has been restricted due to differences in the availability of high-speed internet, especially in underserved and rural areas.
- *Education*: The availability of educational resources, virtual classrooms, and online learning platforms has increased the number of educational opportunities. Still, a lot of places struggle with dependable internet access for pupils.
- *Agriculture*: Precision farming, which makes use of remote sensing and data analytics, may boost agricultural production while lessening its negative effects on the environment. But it's common for farmers in isolated places to lack the connectivity they need to use these technologies.
- *Manufacturing*: The fourth industrial revolution is being propelled by robots, automation, and industrial IoT. Fast networks are necessary for controlling manufacturing operations and exchanging data in real time.
- *Financial services*: These days, it's normal to use mobile payments, online banking, and financial data analytics. In places with little internet connectivity, financial inclusion is still difficult to achieve.

Function of government policy

In order to ensure that everyone has fair access to high-speed internet and to close the digital divide, government policies are essential. Important measures include funding digital literacy initiatives, offering subsidies for internet services, and investing in infrastructure. Policies that promote competition among internet service providers can also aid in lowering costs and raising the caliber of services.

Governments can also encourage innovation by establishing legislative frameworks that facilitate the creation and application of cutting-edge technologies like satellite internet and 5G. Governments can contribute to ensuring that the advantages of the digital age are enjoyed by all by emphasizing digital infrastructure as a crucial element of economic development.

Overcoming the digital gap: the obstacle of remote connectivity

In rural places, there is a noticeable difference between those who have access to high-speed internet and those who do not, known as the digital divide. These areas' remote locations, small populations, and financial limitations have made it difficult to install broadband infrastructure.

The effects of rural connectivity on the economy

There are significant economic ramifications when rural communities lack high-speed internet. Precision agriculture is one example of a technology that farmers and agricultural businesses cannot use to maximize crop yields, minimize resource consumption, and increase profitability. Accessing markets, conducting e-commerce, and offering vital services to clients are difficult tasks for small enterprises in rural areas.

Moreover, attempts to diversify the economy may be hampered by a lack of connectivity. Rural communities frequently depend on a small number of core sectors, like mining or agriculture, which are prone to changes in the market. For people living in rural areas, high-speed internet can facilitate the growth of new enterprises like remote employment or ecotourism, giving them additional sources of income.

Infrastructure difficulties

There are particular difficulties when deploying broadband infrastructure in rural locations. The cost of creating and maintaining networks is increased by the great distances between population centers and the challenging topography. Additionally, internet service providers find it challenging to financially justify the necessary investment in many rural locations due to the low population density.

We need creative solutions to deal with these issues. Even though it's still in its infancy, satellite internet has the potential to give faraway locations broadband access. Point-to-point and point-to-multipoint networks are examples of fixed wireless technologies that can be useful in closing the digital divide. Public-private partnerships and government subsidies may contribute to the affordability and accessibility of these technologies.

The effect on society

Rural communities' inability to access high-speed internet has serious societal repercussions. There are challenges for students attending rural schools when it comes to using educational resources and engaging in virtual learning. Connectivity issues restrict telemedicine services, which have the potential to increase access to healthcare in isolated areas.

Additionally, feeling cut off from society and socially isolated can be exacerbated by the digital divide. Having access to social media and online communication tools can improve community relationships and lessen feelings of loneliness.

It is crucial to close the digital divide in rural areas in order to advance social justice, economic development, and better living standards. Governments, businesses, and communities can collaborate to build a more inclusive and connected future by funding infrastructure, fostering new technology, and putting in place sensible laws.

Case study on the success of rural broadband

The Tennessee City of Chattanooga case

Tennessee's Chattanooga has been a role model for the installation of municipal broadband. The city made significant investments in fiber-optic infrastructure, giving businesses and citizens affordable access to high-speed internet. This program has enhanced educational results, drawn in new enterprises, and boosted economic growth. The example of Chattanooga shows that public ownership and investment in broadband infrastructure are feasible.

Function of satellite internet

Satellite internet has become a practical option in areas with low population density or extremely difficult terrain. Businesses that are substantially investing in this technology include SpaceX, with its Starlink constellation. Satellite internet has always been linked to slower speeds and high latency; however, recent developments have greatly increased performance. Satellite internet can be a lifesaver for towns with spotty broadband connectivity, even though it's not a perfect solution.

Networks of community-owned broadband are an additional strategy to close the digital gap. In order to bring internet services to underdeveloped communities, cooperatives and non-profit groups frequently create these networks. These networks can give members' demands top priority and provide reasonably priced services since they function as cooperatives.

The implementation of broadband in rural areas is greatly aided by government rules and initiatives. Private investment in rural areas can be stimulated by tax rebates, infrastructure development money, and incentive schemes. Broadband network expansion can also be sped up by lowering regulatory barriers and simplifying approval procedures.

Even though these case studies provide insightful information, it's important to understand that rural broadband difficulties are complex. The adoption of broadband services is also influenced by variables including cost, digital content availability, and literacy.

A comprehensive strategy that incorporates infrastructure development, digital skills training, and reasonably priced service plans is needed to overcome these obstacles. Governments, ISPs, and communities can jointly build a future in which all people have access to the advantages of high-speed internet.

RADIATION REDUCING MATERIALS

A defense against the invisible

A constant force in our surroundings, radiation has both positive and negative effects. Although it has uses in industry, research, and medicine, prolonged exposure can have serious negative effects on health. Therefore, the creation of materials that effectively reduce radiation is essential to protecting both the environment and human health.

Recognizing the many forms of radiation is crucial before working with materials. As already discussed earlier, ionizing and non-ionizing radiation are the two main categories into which radiation falls. Ionizing radiation, which includes X-rays, beta, gamma, and alpha particles, has enough energy to strip electrons from atoms. Radio waves and microwaves are examples of non-ionizing radiation, which have lower energy and are typically less dangerous.

The main component of radiation shielding has always been lead. It absorbs X-rays and gamma rays very well because of its high atomic number and density. Products containing lead, like sheets, blocks, and aprons, are frequently utilized in nuclear, industrial, and medical contexts. Lead's weight and toxicity, however, restrict its uses.

Composite materials have come into existence to overcome these restrictions. They give increased flexibility, decreased weight, and occasionally even greater shielding characteristics by combining lead with other elements like tungsten, bismuth, or titanium.

Due to environmental concerns and the demand for lighter, more flexible shielding solutions, the quest for lead substitutes has become more intense.

- *Polymeric materials*: Flexible and lightweight radiation-shielding materials [5] can be made by incorporating heavy metals like bismuth or tungsten into polymers. These composites are used in aircraft, defense, and medical imaging.
- *Nanomaterials*: With their special qualities, nanomaterials like graphene and carbon nanotubes are attractive options for radiation shielding. Their high electron density and aspect ratio can increase the efficiency of shielding.
- *Cement and concrete*: These easily accessible and reasonably priced materials can be strengthened with heavy metals to increase their

radiation-shielding properties. Storage facilities and nuclear power plants frequently employ this strategy.

- *Water*: Water may efficiently attenuate radiation, especially when it's used in large quantities. Spent fuel pools and nuclear reactor cooling systems both make use of this idea.

Creating materials that effectively reduce radiation is a difficult undertaking. Careful consideration must be given to elements including shielding effectiveness, weight, cost, and adaptability. The particular kind of radiation that needs to be shielded also affects the choice of material.

In order to find new materials and enhance those that already exist, research is continuing. The development process is moving more quickly thanks to the combination of sophisticated characterization methods and computational modeling. Furthermore, there's a rising interest in creating multipurpose materials with features like electrical conductivity or thermal insulation in addition to radiation shielding.

Materials that reduce radiation are used in a variety of fields:

- *Healthcare*: Nuclear medicine, patient safety, X-ray, and radiation equipment.
- *Nuclear industry*: Waste management, storage of spent fuel, and reactor components.
- *Aerospace*: Protecting astronauts from radiation in space.
- *Defense*: Keeping people and property safe from nuclear harm.
- *Consumer products*: Appliances, electronics, and building supplies with components that lower radiation.

Materials that reduce radiation are essential for safeguarding both the environment and public health. Although lead has historically been the preferred material, new materials are opening up new avenues for sustainable and efficient radiation shielding. We may anticipate seeing increasingly more advanced and specialized materials emerge in this space as technology progresses.

IMPROVING QUALITY OF LIFE

For both individuals and societies, achieving quality of life—a complex notion that includes physical, mental, social, and environmental well-being—is a basic goal. A harmonious balance of multiple essential components is frequently indicative of a high quality of life:

- *Physical health*: This covers things like having access to good food, exercise, a clean environment, and high-quality healthcare. Preventive care, a balanced diet, and regular exercise are crucial.

- *Mental health*: It's important to have resilience, emotional stability, and the capacity to handle life's obstacles. The availability of mental health services, methods for managing stress, and support networks are important.
- *Social well-being*: Contentment with life as a whole is influenced by a strong feeling of belonging, community involvement, and strong connections. Volunteering, civic engagement, and social support networks are essential.
- *Environmental quality*: Both physical and mental health depend on a clean, healthy environment. The availability of green places, the quality of the air and water, and sustainable living techniques are important.
- *Economic security*: A high quality of life is largely dependent on having stable finances and chances for both professional and personal development. It is crucial to have access to affordable housing, work, and education.

The quest for a great quality of life might be impeded by a multitude of issues. Among them are:

- *Inequality*: Differences in income, access to healthcare, and education pose serious problems.
- *Environmental degradation*: Pollution, resource depletion, and climate change pose a threat to human health and welfare.
- *Social isolation*: A lack of social connections and loneliness can have a detrimental effect on one's general sense of well-being and mental health.
- *Over-reliance on technology*: Digital addiction and excessive screen time can be detrimental to one's physical and emotional well-being.
- *Work-life balance*: The pressures of contemporary life frequently result in stress, burnout, and disregard for one's own well-being.

Improving living quality necessitates a multifaceted, all-encompassing strategy. Some important tactics are as follows:

- *Encouraging physical and mental health*: It is crucial to make healthcare investments, encourage healthy lifestyles, and lower stress levels.
- *Building strong communities*: It's important to encourage social interaction, encourage community involvement, and create inclusive environments.
- Preserving natural resources, investing in renewable energy, and implementing sustainable behaviors are essential for protecting the environment.

- *Reducing economic inequality*: Providing affordable housing, enacting equitable tax laws, and guaranteeing access to high-quality education are crucial.
- *Leveraging technology*: It's critical to minimize the bad effects while maximizing the positive effects of technology use in healthcare, education, and communication.

Achieving a good standard of living is a difficult and continuous endeavor. It necessitates a comprehensive strategy that takes into account how different elements are interrelated. We can improve the future for people and societies all around the world by emphasizing on physical and mental health, fostering strong communities, preserving the environment, addressing inequity, and utilizing technology.

REFERENCES

1. Al-Fuqaha, A., Guizani, M., Mohammadi, M., Aledhari, M., & Ayyash, M. (2015). Internet of things: a survey on enabling technologies, protocols, and applications. *IEEE Communications Surveys & Tutorials, 17*(4), 2347–2376. https://doi.org/10.1109/comst.2015.2444095
2. Fagnant, D. J., & Kockelman, K. (2015). Preparing a nation for autonomous vehicles: opportunities, barriers and policy recommendations. *Transportation Research. Part A, Policy and Practice, 77*, 167–181. https://doi.org/10.1016/j.tra.2015.04.003
3. Norris, J. R., Allen, R. J., Evan, A. T., Zelinka, M. D., O'Dell, C. W., & Klein, S. A. (2016). Evidence for climate change in the satellite cloud record. *Nature, 536*(7614), 72–75. https://doi.org/10.1038/nature18273
4. Liu, Y., Li, Q., & Zhang, Z. (2022). Do smart cities restrict the carbon emission intensity of enterprises? Evidence from a quasi-natural experiment in China. *Energies, 15*(15), 5527. https://doi.org/10.3390/en15155527
5. Garnett, E., & Yang, P. (2010). Light trapping in silicon nanowire solar cells. *Nano Letters, 10*(3), 1082–1087. https://doi.org/10.1021/nl100161z

Chapter 7

Conclusion

INTRODUCTION

The rich tapestry of modern life is clearly woven with technological advances. From the beginning of the digital age to the present, our world has changed dramatically, driven by an insatiable desire for connectedness, convenience, and information. However, the persistent quest for advancement has resulted in an exponential growth in electromagnetic radiation.

Our research into electromagnetic emissions has revealed a complex relationship between innovation and its environmental and biological consequences. The widespread use of wireless gadgets, from smartphones to smart homes, has altered our lives, but it has also presented new challenges. The electromagnetic spectrum, once broad, is becoming crowded, with diverse frequencies competing for space. This congestion affects both human health and the environment. The human body, a wonder of biological ingenuity, is incredibly sensitive to its surroundings. While we have evolved to survive with natural electromagnetic fields, the artificial radiation released by our technology infrastructure creates a new stressor. Chronic exposure to these emissions may have long-term consequences, which are still being investigated. While some research indicates a link between electromagnetic radiation and some health problems, definite causal links are frequently elusive. Nonetheless, the precautionary principle requires that we address this situation with care.

The environmental impact of electromagnetic radiation is also concerning. Natural electromagnetic patterns can be altered, disrupting ecosystems and impacting flora and wildlife in ways we don't yet comprehend. Climate change, already a major worldwide concern, may be exacerbated by the energy consumption connected with wireless devices.

The commercial sector, motivated by the unrelenting quest for profit and market share, has played a critical role in the spread of electromagnetic radiation. The rush for greater data rates, combined with the appeal of automation, has resulted in a disregard for the possible consequences. Data storage, an important aspect of the digital age, has also contributed to the problem. As data quantities grow, the energy necessary to store and analyze this information puts an increasing demand on our resources.

To traverse this complex landscape, we must adopt a comprehensive strategy that promotes both technology growth and human well-being. Mindful automation, which uses technology to improve our lives while protecting our health and the environment, is critical. The proper utilization of frequency bands, together with the development of radiation-reducing materials, can assist in reducing the hazards associated with electromagnetic emissions.

Furthermore, we must invest in R&D to investigate alternative energy sources and increase energy efficiency. By reducing our reliance on fossil fuels, we can not only address climate change but also relieve strain on the electrical system, reducing the need for new infrastructure that contributes to electromagnetic pollution.

Ultimately, the survival of our planet and its inhabitants is dependent on our ability to strike a balance between innovation and accountability. By making educated decisions and demanding responsibility from corporations and governments, we can build a world in which technology is a force for good rather than a threat to our health and well-being. It is critical that we move beyond simply accepting the status quo and actively work toward a future in which technology and nature coexist together.

Key areas for future research and action

1. Long-term health consequences of prolonged exposure to electromagnetic radiation.
2. Effects of electromagnetic radiation on ecosystems and biodiversity.
3. Developing more efficient and sustainable energy sources.
4. Development of worldwide guidelines for electromagnetic emissions.
5. Campaigns to raise awareness of technology's risks and benefits, as well as to promote responsible innovation and ethical technology development.

Addressing these difficulties and taking a proactive approach can pave the path for a healthier, more sustainable future.

The impact of electronic devices on human health and society, focusing on the potential health problems connected to excessive gadget use, must be focused upon. Radiation, a result of many electronic equipment, comes from various sources, including natural background radiation and man-made devices like cell phones, Wi-Fi routers, and microwave ovens. Efficient energy use is crucial to meet the increasing demand for electronic gadgets, emphasizing the importance of breakthroughs in battery technology and energy-efficient designs.

The human nervous system's sensitivity to electromagnetic fields (EMFs) produced by wireless communication equipment should be researched.

Chronic exposure to EMFs may have consequences for brain health. The importance of weighing technological benefits against health concerns and providing a thorough review of the relationship between modern technology, radiation exposure, and human health must be noted.

Energy harvesting, capturing ambient electromagnetic energy and converting it into usable electrical power, has the potential to power small electronic devices and sensors, contributing to more sustainable and self-sufficient systems.

Also, the Green energy alternatives, emphasizing the need for sustainable approaches for capturing and exploiting electromagnetic energy, aim to reduce energy consumption's environmental impact by incorporating renewable energy sources and improving energy efficiency.

Regulatory recommendations aim to reduce exposure and prevent negative health consequences. The significant advancements in technology and its impact on the spectrum, highlighting the need for faster data rates and improved life automation, must be monitored. The rapid growth of internet speeds has transformed daily life by automating operations in homes, offices, and businesses, enhancing the economy and convenience but also raising concerns about desensitivity in human connections. The reliance on technology for communication and task management may result in less personal engagement and emotional relationships.

Data storage limits are a major concern, as the exponential expansion of data output requires effective and safe solutions to store massive volumes of information. The investigation into present and future prospects for data storage systems that can keep up with the ever-increasing data needs are sure research problems. Energy cutoffs are another important topic, as the increased use of technology puts a strain on energy resources. Understanding these dynamics allows for insights into managing technology improvements while tackling the difficulties of data storage and energy usage.

Sensible decisions about automating life in an age of rapid technological growth are the need of the hour. It emphasizes the need to negotiate the accompanying frequency chaos, which can cause signal interference and poor performance. Effective management of frequency bands is required to enable smooth and dependable communication.

Limiting emissions from electronic devices and industrial processes is essential to reduce their environmental impact. Balancing the requirement for speed with sustainable methods is critical for long-term success. Radiation-reducing materials in the design and manufacture of electronic equipment can reduce the detrimental impacts of electromagnetic radiation on human health and the environment, promoting safer and more sustainable technology use.

To improve the quality of life through intelligent automation choices, effective frequency management, emission controls, appropriate data rates, and the use of radiation-reducing materials, it is possible to reap the benefits of technology while reducing its negative consequences.

Smartphone addiction is a growing concern globally, fueled by the constant urge to browse social media. This issue has prompted many, particularly in the USA and Europe, to turn to "dumb phones" as a potential solution. A "dumb phone" is a device with limited functionalities, primarily enabling calls, text messages, and basic navigation, devoid of the advanced features and apps found in modern smartphones.

A study by the National Institute of Health in the USA reveals that 74.3% of smartphone users feel dependent on their devices. This dependence has significant mental health implications, including stress, depression, and anxiety. The perpetual need to stay connected and receive notifications exacerbates these issues, particularly among young people who experience FOMO (fear of missing out). A Harvard University research further highlights that social media apps trigger brain responses similar to those caused by addictive substances.

"Dumb phones" offer a way to mitigate these adverse effects by reducing screen time and social media usage. Unlike feature phones, which might still have access to social media, true "dumb phones" focus solely on essential communication and navigation. This makes them particularly suitable for young users, as early exposure to smartphones can lead to developmental issues and addiction.

Technology is the architect of progress, a catalyst for innovation, and a bridge to a world once unimaginable. In the realm of social media, it has forged connections, ignited movements, and amplified voices. It is a marketplace of ideas, a platform for creativity, and a tool for empowerment. When wielded with wisdom, it can be a beacon of hope, illuminating paths to a better future.

Yet, the same technology, when misused, can become a weapon of mass destruction, capable of eroding societal fabric, fracturing relationships, and poisoning minds. The allure of the digital world can ensnare us, isolating us from genuine human connection, distorting our perception of reality, and feeding into a vortex of negativity. The misuse of social media can breed division, hatred, and misinformation, threatening the very foundations of democracy and civility.

It is imperative that we harness the power of technology and social media as a force for good. We must cultivate digital literacy, teaching ourselves and future generations to navigate this complex landscape with discernment and critical thinking. We must prioritize authenticity, empathy, and compassion in our online interactions, fostering a culture of respect and understanding.

Moreover, we must strive for balance. Technology should be a tool, not a master. It should enhance our lives, not consume them. Let us use it to connect, to learn, to grow, and to inspire. Let us create a digital world that reflects the best of humanity, a world where technology is a servant to our aspirations, not a dictator of our destinies.

The choice is ours. We can shape technology to serve our collective well-being, or we can allow it to enslave us. The future of humanity hinges on our ability to wield this powerful tool with wisdom and responsibility. Let us strive to be architects of a harmonious coexistence between humanity and technology, a world where innovation and compassion intertwine to create a brighter tomorrow.

Index

For Product Safety Concerns and Information please contact our EU
representative GPSR@taylorandfrancis.com
Taylor & Francis Verlag GmbH, Kaufingerstraße 24, 80331 München, Germany

www.ingramcontent.com/pod-product-compliance
Lightning Source LLC
Chambersburg PA
CBHW072254210326
41458CB00073B/1721